United Nations
Development
Programme

SPECIAL PUBLIC WORKS PROGRAMMES

International
Labour Office

Tree Nurseries

AN ILLUSTRATED TECHNICAL GUIDE AND TRAINING MANUAL

Booklet No. 6

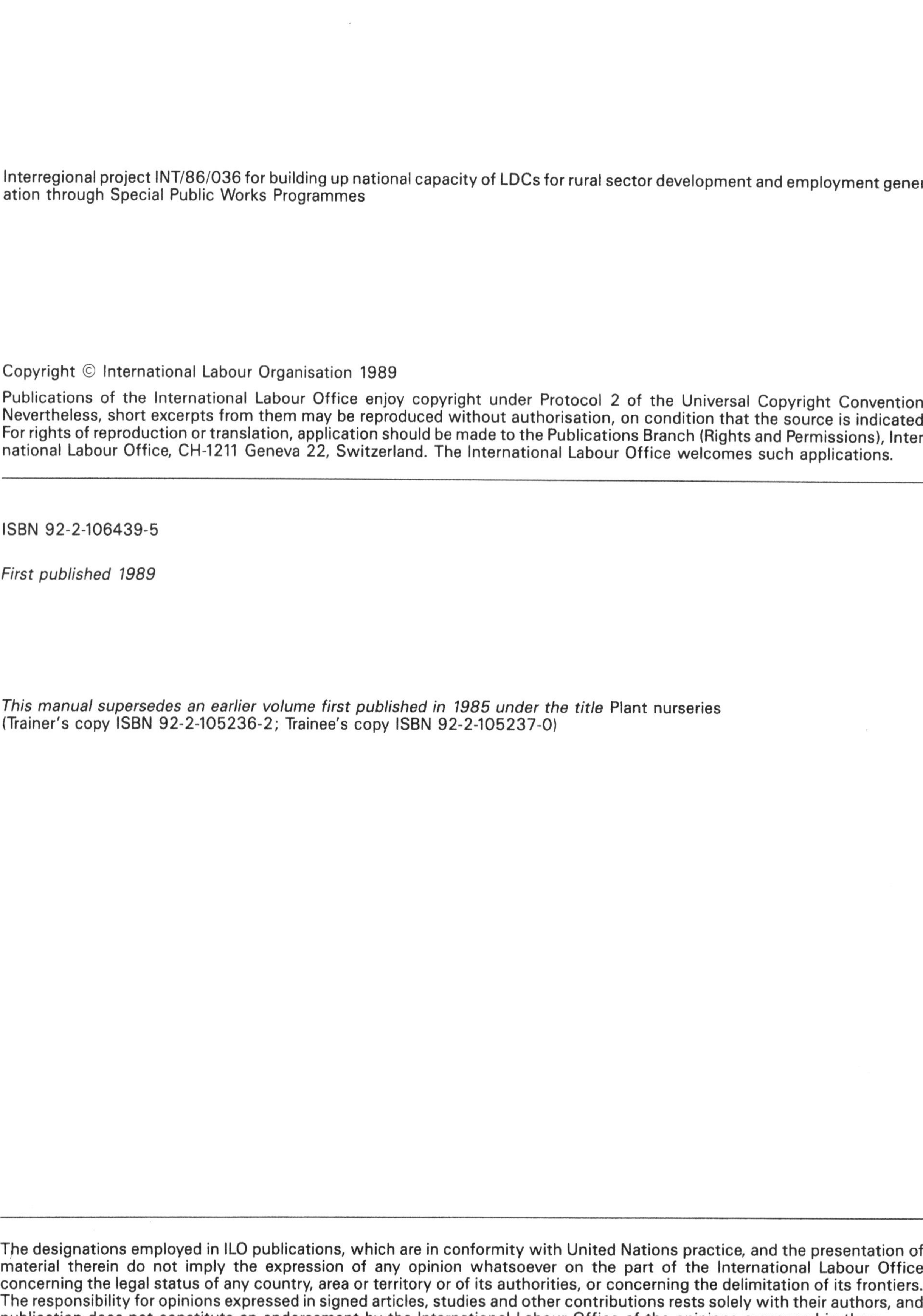

Interregional project INT/86/036 for building up national capacity of LDCs for rural sector development and employment generation through Special Public Works Programmes

ISBN 92-2-106439-5

First published 1989

This manual supersedes an earlier volume first published in 1985 under the title Plant nurseries (Trainer's copy ISBN 92-2-105236-2; Trainee's copy ISBN 92-2-105237-0)

Printed in Switzerland

IDE

CONTENTS

INTRODUCTION

WHAT IS A TREE NURSERY AND WHY IS IT NEEDED?

A nursery is an area where young plants can grow with special care and protection.

It produces seedlings for afforestation and tree planting. Seedlings are usually needed in large numbers and young trees of most species do not survive well if directly grown on the plantation site. It is therefore easier and cheaper to grow seedlings in one place - the nursery - and plant them only when they need less care and protection.

The purpose of a nursery, therefore, is to grow seedlings:

- of the required species;
- of the right size and sturdiness at the beginning of the planting season;
- in sufficient numbers

for the intended tree planting programme.

ABOUT THIS GUIDE

This guide contains all the basic information needed to set up and successfully run nurseries relying on manual work. It focuses on the production of containerised seedlings as the method most commonly used in the ILO's Special Public Works Programmes (SPWP) and other non-industrial tree planting in developing countries.

There are many differences between countries and regions regarding the tree species grown, the climates, the tools available and other factors. All this variety could not be covered in this small booklet and some of the advice may have to be adapted to local conditions.

Experience shows, however, that nurseries have many things in common and that it is often the basic installations, equipment and techniques that lead to poor quality seedling production. These aspects are therefore emphasised. Common errors that one comes across in nurseries are listed at the bottom of each paragraph in the chapter on nursery operations.

This booklet should be considered a starting point upon which the nursery workers' and managers' experience can and must be built, through careful observation. The best guide to nurseries is the experience of the people running them.

The booklet was produced thanks to the financial contribution of the United Nations Development Programme inter-regional project INT/86/036 - "Building up national capacity of LDCs for rural sector development and employment generation through Special Public Works Programmes", as well as a contribution by the International Labour Office. It was written by Peter Poschen (ILO Industrial Activities Branch), illustrated by Anja Laengst (ILO external collaborator), and edited and prepared for printing by Hazel Cecconi (ILO Industrial Activities Branch).

---

1. SETTING UP A TREE NURSERY

1. SETTING UP A TREE NURSERY

1.1 SITING THE NURSERY

When selecting the nursery site, two kinds of factors have to be considered:

(a) FACTORS IMPORTANT FOR SEEDLING GROWTH:

- water
- soil
- topography
- materials.

(b) FACTORS IMPORTANT FOR NURSERY MANAGEMENT:

- location
- access
- ownership
- size.

(a) FACTORS IMPORTANT FOR SEEDLING GROWTH

Water:

A guaranteed water supply all year round is absolutely essential. Water is needed most when least available. From 10-20 l of water per day per m^2 of productive area are needed (depending on prevailing temperatures). Sources of water may be: springs, small streams, rivers, ponds, boreholes or wells. The construction of a small channel diverting water into or close to the nursery can greatly reduce time and cost of watering. High salinity which may occur in dry areas (water evaporates leaving a salt crust) should be avoided.

WATER SUPPLY

SURFACE WATER

Creek or river

Lake or pond

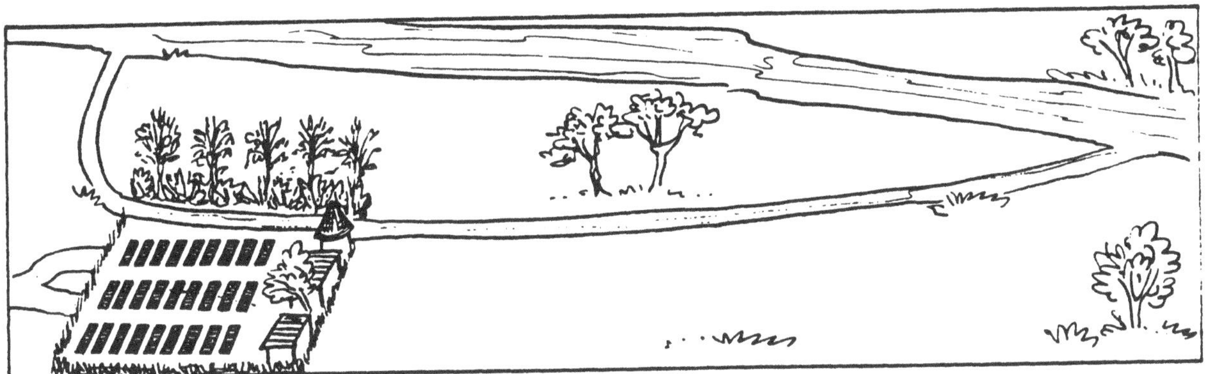

River and channel

GROUND WATER

Water table high

Water table low

1. SETTING UP A TREE NURSERY

1.1 SITING THE NURSERY

Soil:

For potted plants, the soil of the nursery site itself is not so important. However, suitable potting soil should be available close to the nursery. It should not be too sandy because otherwise it falls out of the pots, nor contain too much clay because extreme swelling and shrinking damages seedling roots (you can determine whether soil contains too much clay by rubbing some moist soil between thumb and index finger - it should have the consistency of flour; it should not stick or shine).

The soil should have a good water-holding capacity but adequate drainage.

Soils of extreme acidity or alkalinity/salinity should be avoided: they can usually be recognised by their vegetation cover because it consists of specially adapted plants, while agricultural crops cannot grow on such soils.

For bare-rooted stock see p. 64.

Topography:

The nursery should not be exposed to dessicating winds such as those prevailing on hilltops, nor to flooding or severe frost, as may occur in valley bottoms.

Relatively flat land is most suitable, ideally with 1-2 per cent slope because this permits the water to run off so that no water-logging nor erosion occurs.

If flat land is not available, terraces have to be constructed.

The site should receive sunlight for the major part of the day. Shading may be desirable to avoid excessive heating.

Materials:

To condition soil texture, water-holding capacity, drainage and fertility, it is convenient if sand or fine gravel and humus, forest soil or turf are available close to the nursery.

Poles and sticks should also be available for fencing, shading, buildings, etc.

SOILS TO AVOID

Clay-cracking pattern

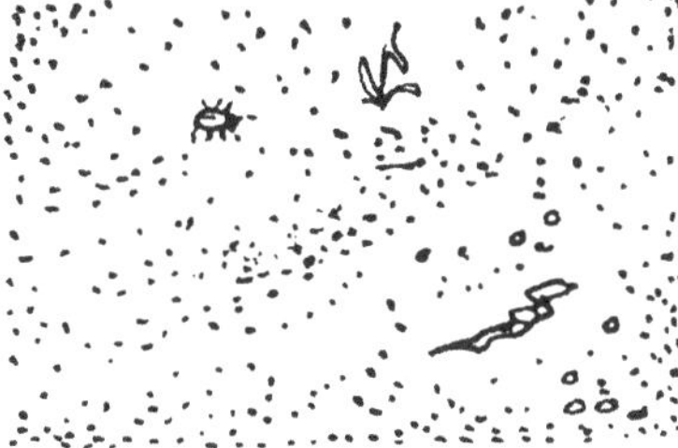

Sand

Stones

TOPOGRAPHY

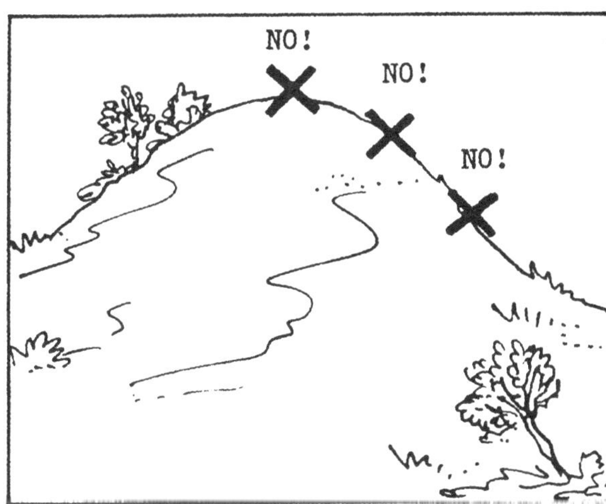

Hilltops/steep slopes

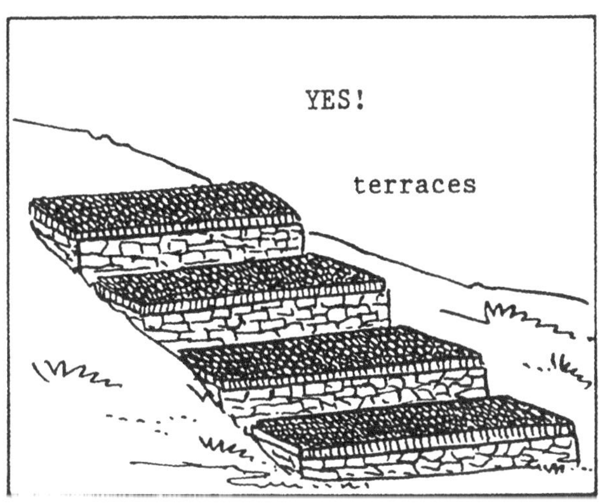

Moderate slopes

Waterlogged or temporarily flooded riverine areas and depressions

1. SETTING UP A TREE NURSERY

1.1 SITING THE NURSERY

(b) FACTORS IMPORTANT FOR NURSERY MANAGEMENT

Location

Close to the planting area to minimise transport.

Access

Near supervisor's living quarters ("The farmer's footprints = best manure for the farm").

Accessible at all times, preferably close to the road.

Ownership

The nursery terrain should preferably be public land (owned by the government or the community) so that management may be free in its decisions. Make sure traditional rights to use the land (e.g. fuelwood collection, grazing or water) do not interfere. For private land and water rights, clear lease agreements have to be made.

An alternative may be to assist individuals or groups (farmers, women, schools, etc.) to set up a nursery, for instance within a private garden.

Size

The size of the nursery area depends on:

- the number of seedlings required for planting;
- the time it takes to produce seedlings of the desired size and species in the nursery;
- the size of the containers used.

The prototype layout shown opposite is for a medium-sized nursery with an annual capacity of about 100,000 containerised plants using pots of 8 cm diameter when filled. (If pots of, say, 4.5 cm diameter filled are used, 250,000 seedlings can be accommodated.) This layout can be scaled up or down according to the specific requirements. Appendix 3 provides a table helpful for calculating space requirements.

LAYOUT OF NURSERY

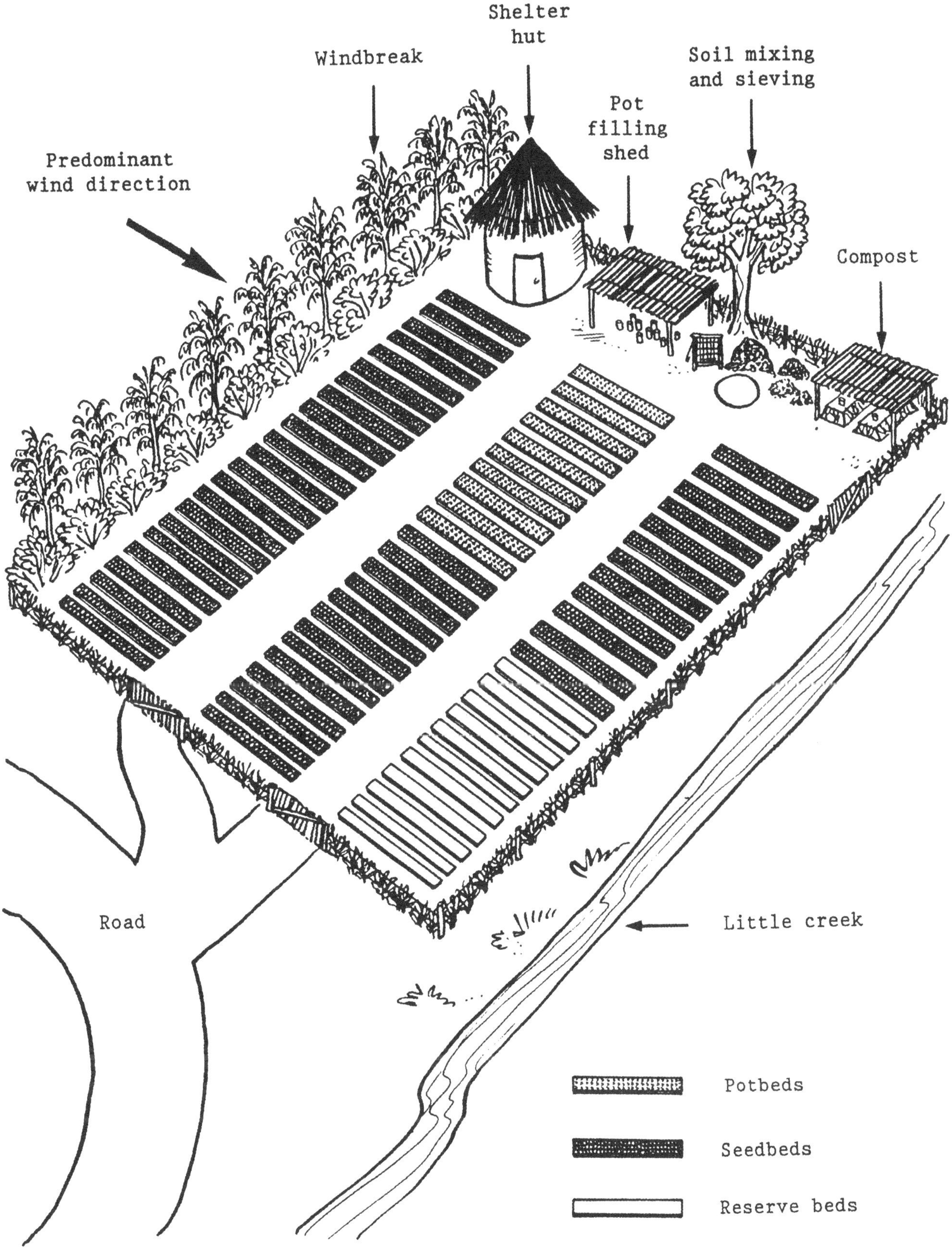

Shelter hut

Windbreak

Soil mixing and sieving

Pot filling shed

Predominant wind direction

Compost

Road

Little creek

Potbeds

Seedbeds

Reserve beds

1. SETTING UP A TREE NURSERY

1.2 BASIC FACILITIES

A nursery must have:

(a) access roads and paths;

(b) fencing;

(c) a shelter for tools, materials and workers;

(d) a soil dump;

(e) seedbeds;

(f) potbeds;

(g) shading and shelter for plants.

In addition, it may have:

(h) windbreaks;

(i) a compost area;

(j) a working shed;

(k) a seed extraction area.

(a) Access roads and paths

The siting of the nursery should minimise the need for transport of materials and seedlings with vehicles. Where seedling delivery relies on transport with vehicles, access ways have to be wide enough and allow for turning, particularly if trailers are to be used. The path/road surface has to be sufficiently stable to support the vehicles during the rainy season.

(b) Fencing

Protection is needed against browsing and trampling animals.

Initially thorny branches and poles should be used. Fences have to be at least 1.5 m high.

If the nursery is to be used for a longer time, living fences/hedges are advantageous protection against wind, animals, and interception of weed seeds. Plant shrubs and trees with dense branches down to the ground, e.g. Cypress, Lantana. Species easily propagated by cutting are very convenient. Where stones are more readily available than other materials, stone walls are a convenient means of protection.

FENCES

DEAD FENCING

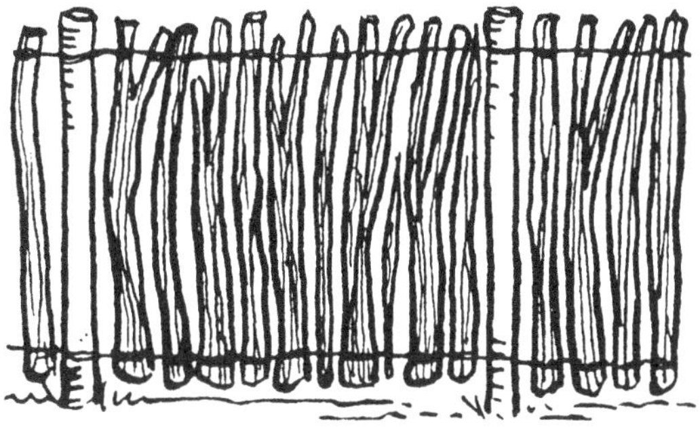

1.50m

Sticks

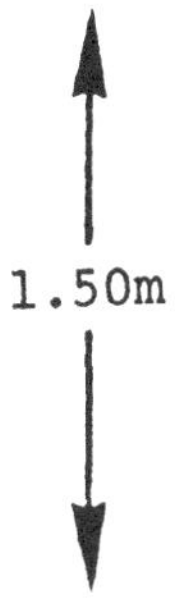

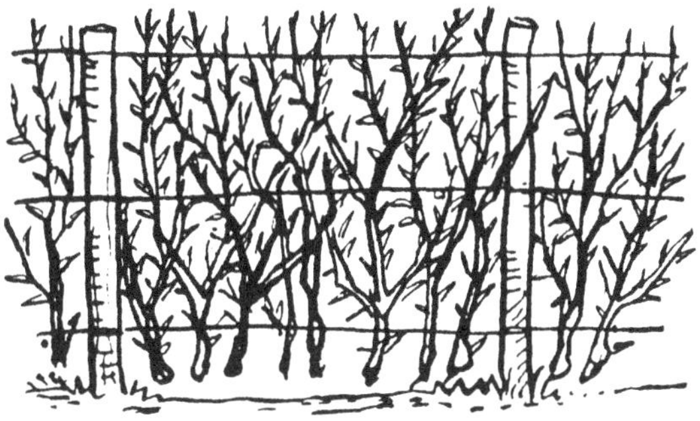

Thorny branches

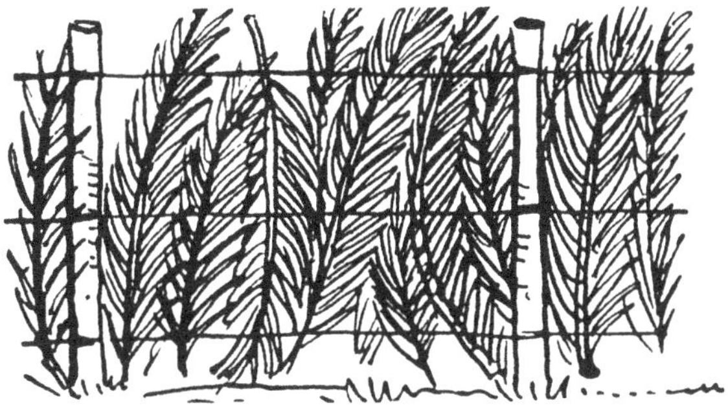

1.50m

Palm leaves

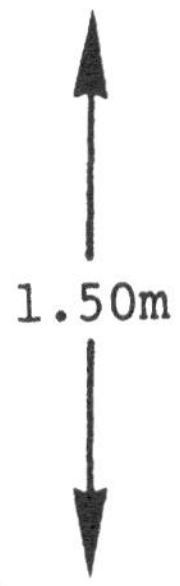

Stone wall

LIVE FENCING

Bamboo fence combined with hedge

Hedge of coniferous trees

1. SETTING UP A TREE NURSERY

1.2 BASIC FACILITIES

(c) A shelter for tools, materials and workers

As a shelter for the workers and to keep tools and materials in a dry and safe place, a simple, small hut should be constructed from materials locally available (poles, bamboo, mud, etc.) in the local style.

(d) A soil dump

The production of potted seedlings requires large amounts of soil, sand and humus.

Space is needed to store these materials separately next to a place where they can be mixed and the pots filled.

(e) Seedbeds

Seedbeds should be 1 m wide with a 60 cm path between them. The bed surface should be level with a low earth embankment along the margin of the bed to prevent the irrigation water from running off and washing away soil, seed or mulch.

Best germination results are obtained if the seeds can readily absorb water, easily extend their roots and break the surface of the bed with little resistance. A germinating seed must not dry up even temporarily because this would kill it. Stagnant water and permanent high humidity, on the other hand, increase the risk of diseases, particularly damping-off. To obtain optimum conditions, the ideal seedbed consists of the three layers shown in the figure opposite:

(i) gravel to ensure good drainage;

(ii) humus-rich soil holding water;

(iii) soil and sand mixture easy to penetrate and well drained.

The soil in seedbeds should be changed after some years to avoid diseases (it may be used to fill pots).

In high rainfall areas where waterlogging occurs frequently, raised seedbeds should be used.

MIXING AND SIEVING POTTING SOIL

RAISED SEEDBED

Earth mound

Path

60cm

100cm

Boards if available

5-10cm mixture sand + soil 1:1

5cm humus-rich soil

5cm gravel

1. SETTING UP A TREE NURSERY

1.2 BASIC FACILITIES

(f) Potbeds

Beds 1 m wide and 5-10 m long are most convenient. If the beds are wider, weeding and watering the centre is more difficult. On terraces, beds must be along the contour.

Between the beds, access paths are needed for transport, weeding, watering, etc. They should preferably be 60 cm wide to permit the use of wheelbarrows. Note that paths are needed on either side of the beds. On flat land, this is achieved with one path per bed. On terraces, additional paths are required.

Normally, seedbeds are level with the ground.

In high rainfall areas and where waterlogging occurs frequently, raised potbeds should be used.

In dry, hot climates and on well-drained soils (sands, gravel), sunken seedbeds are convenient to protect the pots from excessive heating.

To keep the pots in an upright position, frames of local materials like stones, bamboo, poles, sawn timber, bricks, rope or simply earth should be placed on all sides of the potbeds. Solid frames made from earth or sawn timber have the additional advantage of reducing heating and the growth of algae in the pots of the outer rows.

POSITION OF POTBEDS

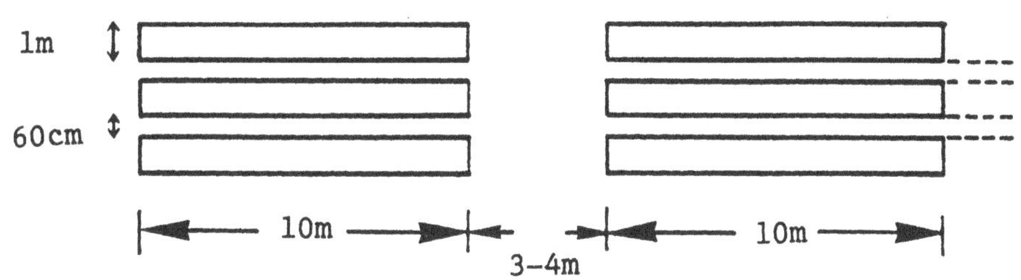

TYPES OF POTBEDS

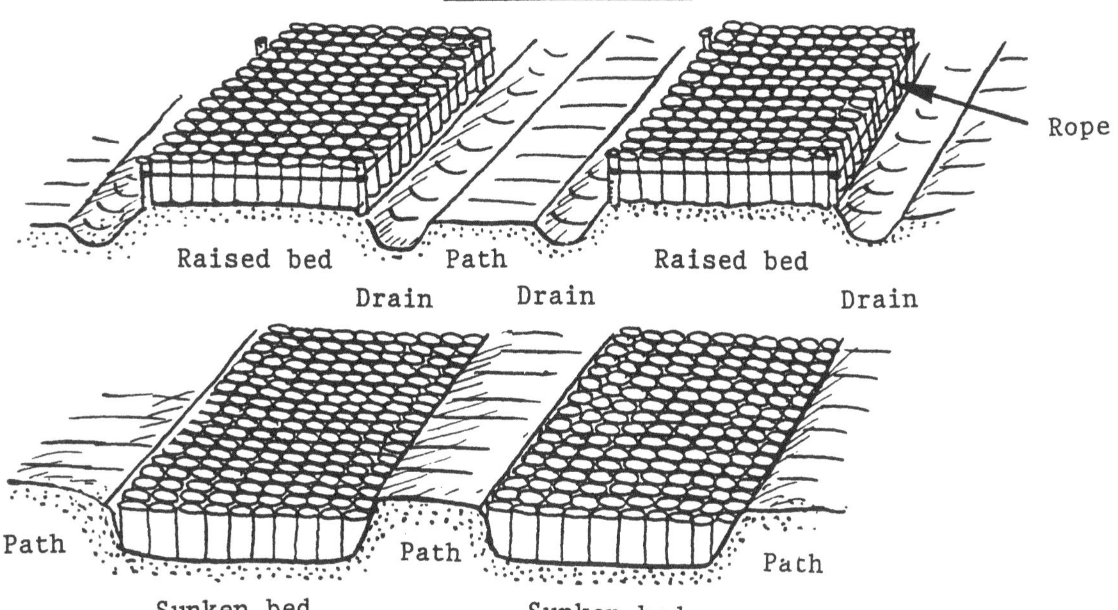

TYPES OF FRAMES

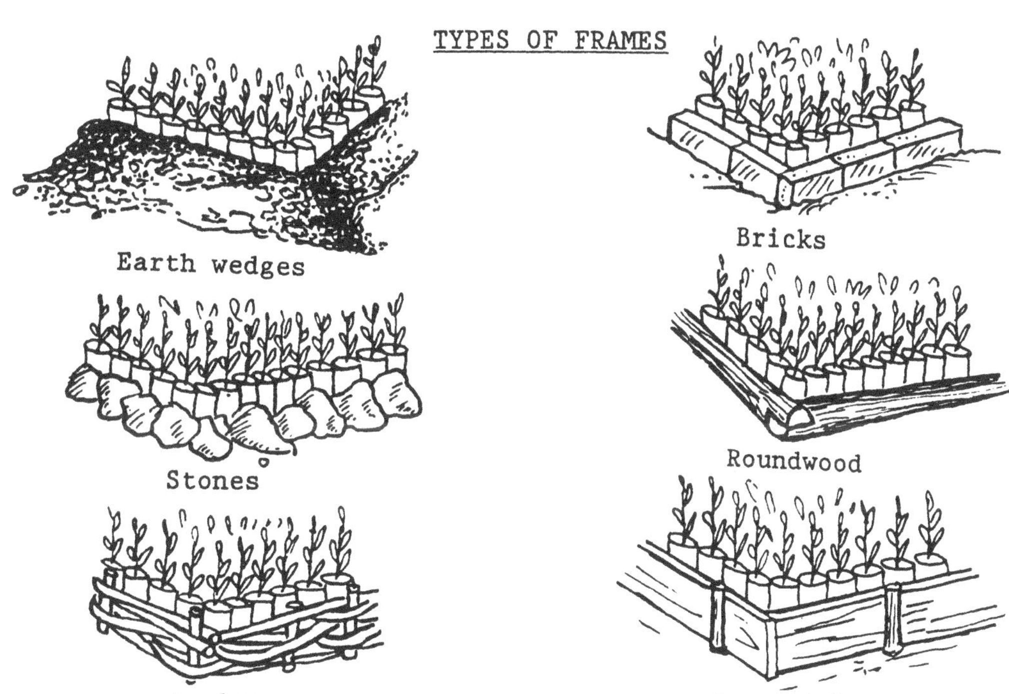

1. SETTING UP A TREE NURSERY

1.2 BASIC FACILITIES

(g) Shading and shelter from frost, hail and heavy rain

All seedlings need shelter from the sun when they are still very young and delicate. Shades are therefore made from locally-available material such as sorghum or millet stalks, bamboo, grass, etc. Because shading is only temporary, they should be easily removable. The best method is to tie the material with ropes so that the shades can be rolled when not needed. The mats can conveniently be placed on frames made of branchwood, stalks or wire of a height of about 50 cm.

Shading mats should allow about half the sunlight to pass through. If they are too dense, seedlings will grow too slowly. Where frost, hail or heavy rain occur, especially dense mats or a double layer of normal shading mats can be used at night and during storms to protect seedlings.

(h) Windbreak

Where strong and/or dessicating winds prevail, a windbreak consisting of at least one, but preferably two, row(s) of shrubs and one row of trees should be planted along the nursery side facing the main wind direction. Larger species with medium dense crowns are best, e.g. *Filaho*, *Cassia*, *Eucalyptus*, *Grevillea*. The trees should be fast-growing in order to become quickly effective.

The trees should be far enough from the beds to avoid crowns extending over them, even when trees have grown bigger.

In some cases, windbreaks made of dead material may be needed permanently or as a temporary solution until living windbreaks become effective.

POT BED SHADING

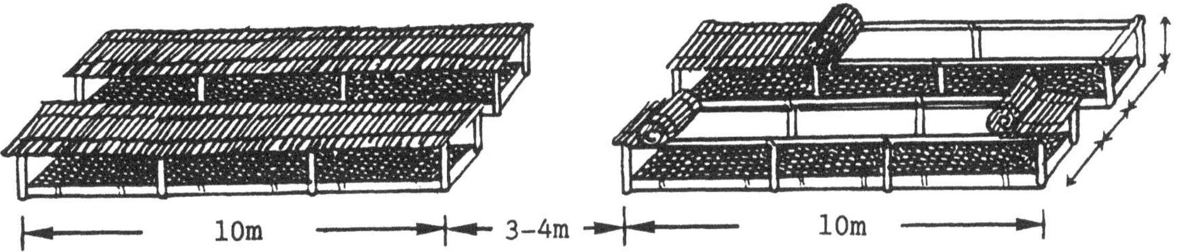

Standard pot beds with removable shades on frames built from branchwood

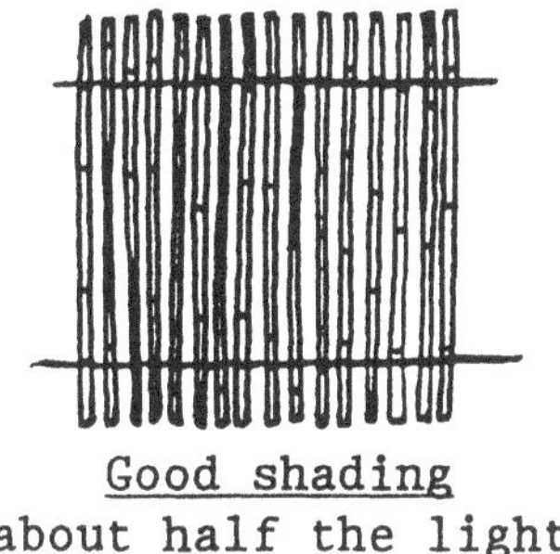

Good shading
about half the light can penetrate

Bad shading
mat too thick for shading but useful as protection against frost, heavy hail or rainstorms

WINDBREAKS

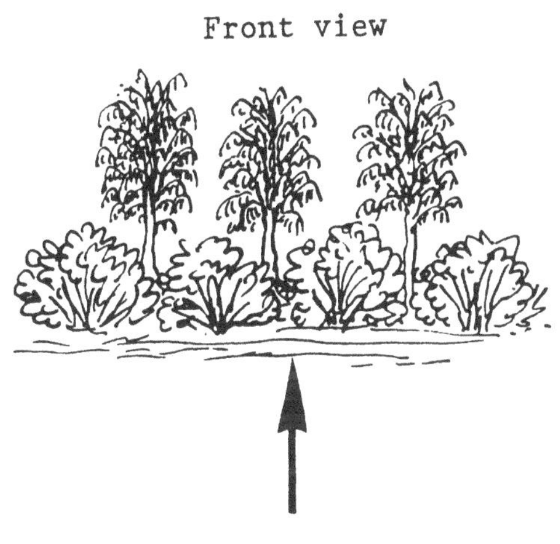

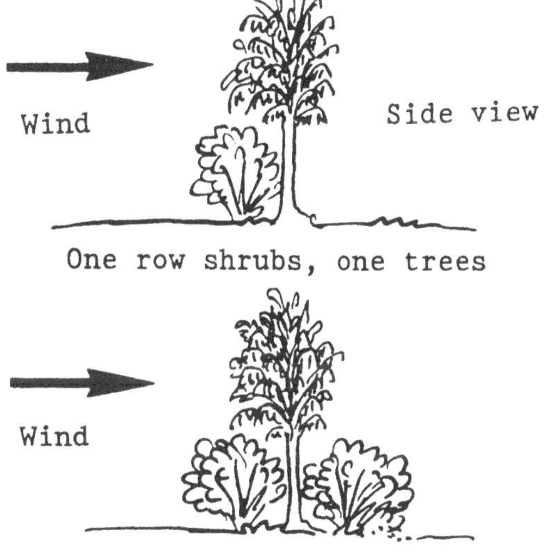

1. SETTING UP A TREE NURSERY

1.2 BASIC FACILITIES

(i) Compost

Heaps

In humid climates, compost is prepared in heaps/piles because compost pits would fill with water, interrupting the formation of compost.
The pile may be 1-1.30 m high, 2 m wide and 2-4 m long.

Pits

Compost pits are dug in drier climates. They are 1-1.5 m deep, 1.0 m wide and 2 m long.

For both compost pits and compost heaps, a shading mat/roof has to be constructed, unless there is shade from trees.
Enough space has to be available next to the compost pile or pit to turn the compost over while it is maturing. (For more details on compost, see Appendix 1).
Compost heaps and pits are not places for rubbish disposal! (No plastic bags, tins, bottles, etc.)

(j) A working shed

To protect workers, potting soil and plants from sun and rain during pot filling and pricking out of plants from seedbeds to pots, a shed should be erected. To accommodate workers and pots awaiting transport to the potbeds a shed 3 x 5 m wide and 2 m high is appropriate for a nursery with a capacity of 100,000 plants/year.

(k) A seed extraction area

If the project collects the seed it needs, an area of 10 m^2 for drying and extraction of seed from pods, cones, etc. has to be available. This area should receive full sunlight to reduce drying time. It should be level. For drying and extraction, seeds are placed on mats or canvas to avoid mixing with soil, weeds, etc.

COMPOST PILES

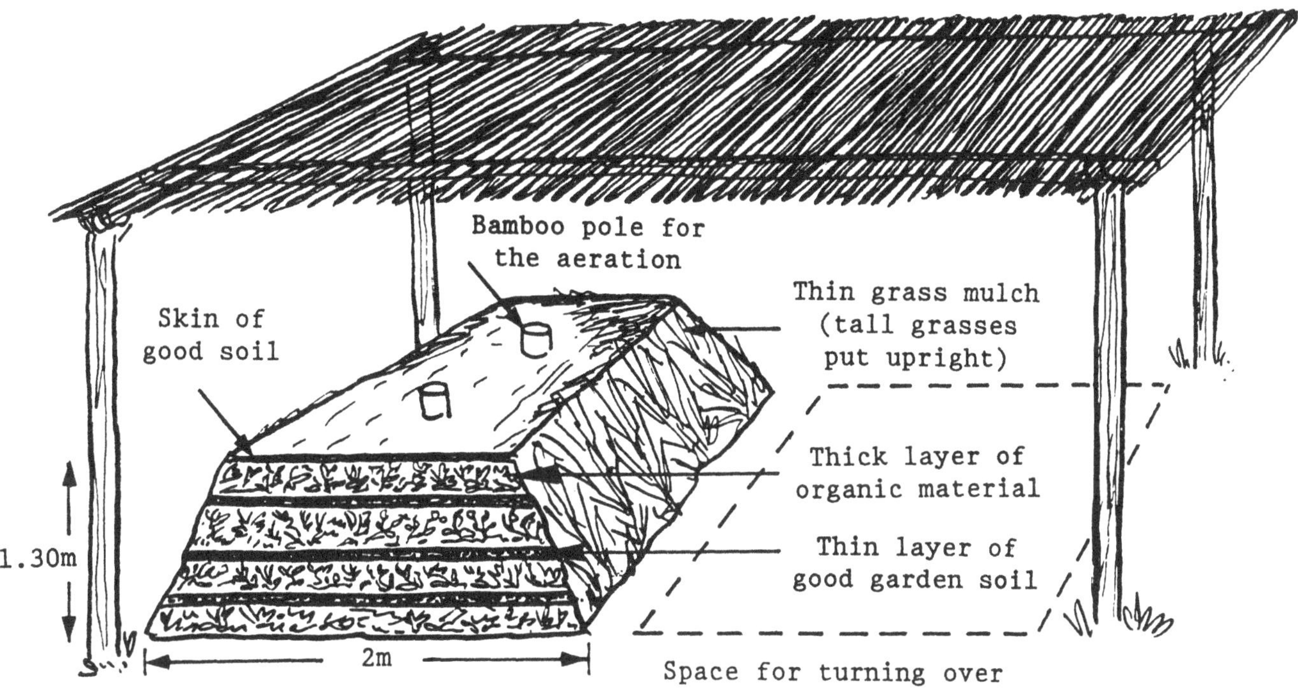

COMPOST PITS

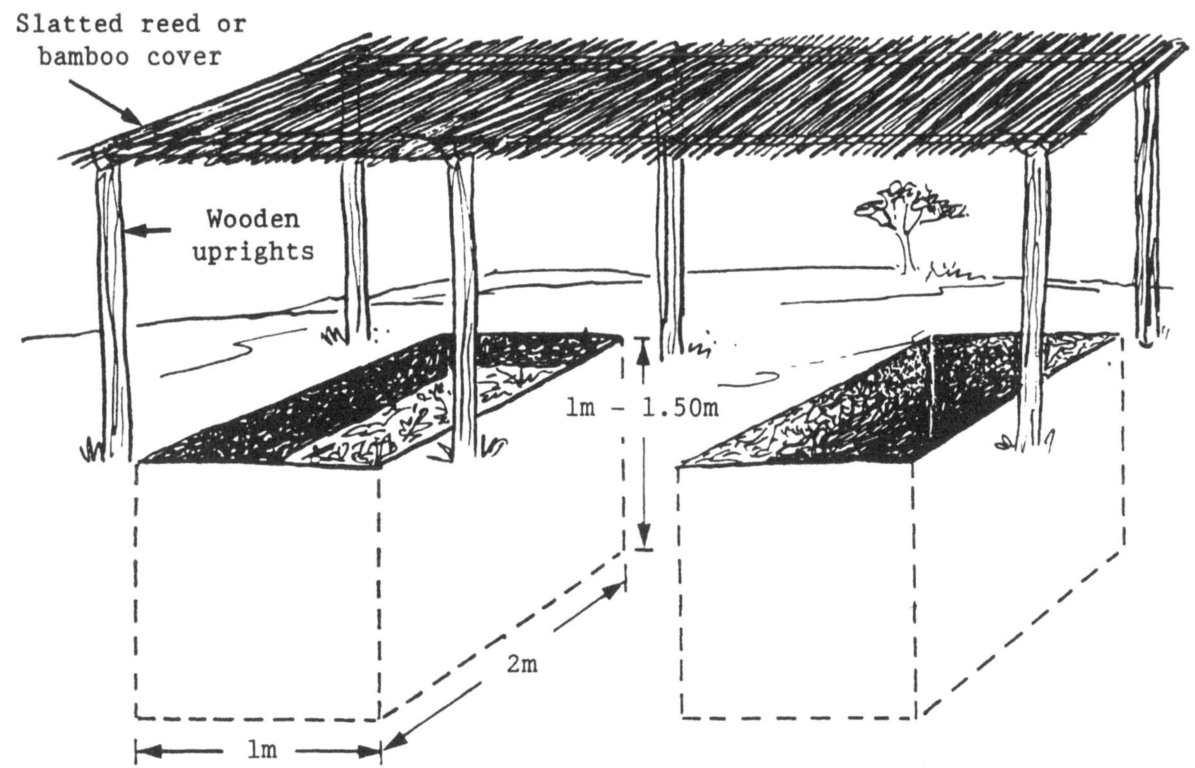

1. SETTING UP A TREE NURSERY

1.3 NURSERY LAYOUT

Shape: The nursery should, if possible, be square shaped to minimise boundary lines which have to be fenced, and to avoid unnecessary transport. If the terrain does not permit this, it should be of rectangular shape.

Size: When determining the required size, the area needed for production (potbeds, seedbeds and reserve beds) and for roads, paths, buildings, fences, windbreaks, etc. has to be taken into account. An area for possible extension in the future may also be desirable.

Example:

For the following example, we assume that 85,000 seedlings are required per annum, including replacement of last year's failures.

Because not all plants raised will survive to seedling stage and a certain number will not be of good enough quality to be delivered for planting, more plants have to be produced than will be planted. This reserve should be about 10-15 per cent of the requirements. For 85,000 seedlings, this is, say, 11,000.

We assume that pots of 20 x 12.5 cm flat are used. When filled, their diameter is about 8 cm. One square metre of potbed can hold 156 pots of this size. 96,000 pots require:

$$\frac{96,000}{156}\ m^2 = 615\ m^2 \text{ of potbeds}$$

NURSERY LAYOUT 1:300

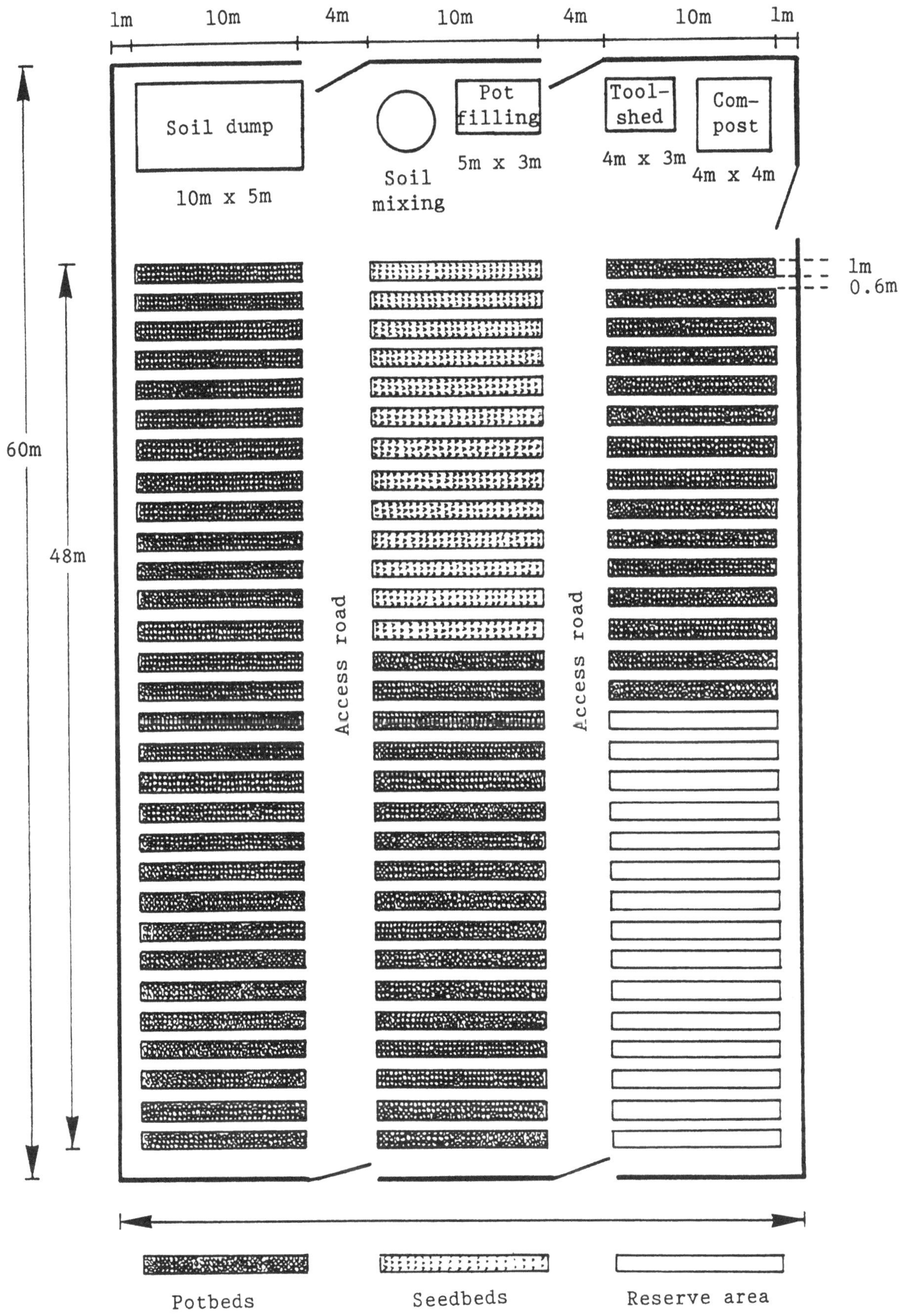

1. SETTING UP A TREE NURSERY

1.3 NURSERY LAYOUT

The seedbed area can be estimated as 20 per cent of the potbed area. For our example this would be: $625 \times 0.2 = 125\ m^2$.

An additional 20 per cent of the bed area is set aside as a reserve area to produce a larger number of plants or seedlings in larger containers (e.g. for difficult sites or fruit tree seedlings for grafting). This brings the total bed area to 900 m^2.

Preferably, all seedbeds and potbeds should be of the same size and shape. The roads should be wide enough to permit passage of pick-up cars and light trucks, i.e. 3-5 m, preferably 4 m. Indications for the dimensions of buildings and other facilities are given in the table opposite. As can be seen, more non-productive than productive area is needed. As a rule of thumb, the non-productive area will be twice the productive (i.e. seed/potbed) area.

Example of area required:

Productive area	No. of beds	Size of beds	Area
Seedbeds	13	10 x 1	130 m^2
Potbeds	62	10 x 1	620 m^2
Reserve bed area	15	10 x 1 (20%)	150 m^2
Total productive area			900 m^2

Non-productive area (facilities)		
Access paths between beds	0.6 m wide	600 m^2
Roads	4 m wide	430 m^2
Tool shed	3 x 4 m	12 m^2
Soil dump	5 x 10 m	50 m^2
Soil mixing		10 m^2
Working shed	3 x 5 m	15 m^2
Compost	4 x 4 m	16 m^2
Fences and windbreaks		260 m^2
Vacant		117 m^2
Total non-productive		1,500
Total area		2,400 m^2
		=====

Note that in favourable conditions where survival rates of plants raised in smaller pots are satisfactory, the same bed area can produce a much higher number of seedlings.

1. SETTING UP A TREE NURSERY

1.4 TOOLS REQUIRED

(a) FOR WORKING THE SOIL

Pickaxe	(1)	Is used to break up hard and stony ground. The steel head is double-ended with a point at one end and a blade at the other.
Traditional hoe	(2)	Is used for loosening the soil.
Shovel	(3)	Is used for moving earth, for sieving soil and soil mixing.
Flat-pronged fork	(4)	Is used for loosening soil, to lift bare-rooted seedlings and to turn over compost. Where work is done bare-footed or with light shoes, the fork should have a small foot-rest.
Rake	(5)	Breaks up and levels the soil. It has a row of 10-16 teeth and is fitted with a 1.80 m handle.

(b) FOR LAYOUT

Tracing line	(6)	Thin hemp or nylon cord, 10 m long (with knots at 1 m intervals), attached to 30-50 cm long pegs at each end, used to trace straight lines, e.g. borders of seedbeds or potbeds and to measure distances.

TOOLS REQUIRED

(1) Pickaxe

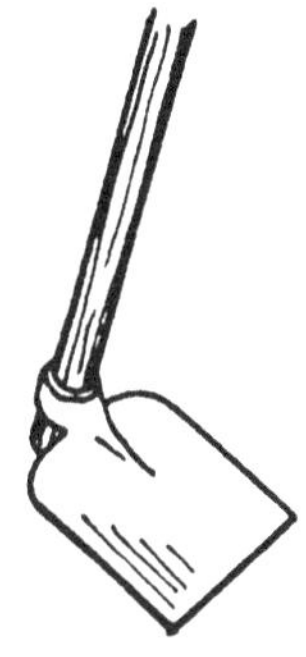

(2) Traditional hoe

(3) Round-nosed shovel

(4) Flat-pronged fork

(5) Rake

(6) Tracing line

1. SETTING UP A TREE NURSERY

1.4 TOOLS REQUIRED

(c) FOR PREPARATION OF POTTING SOIL AND POT FILLING

Sieve (7) Soil for the seedbeds and for potting should not contain larger-size particles, stones, pieces of wood or the like. If necessary, it is therefore passed through a coarse sieve of approx. 1.5 cm (half an inch) mesh wire fitted to a wooden or metal frame of approx. 1 x 1.5 m.

Funnel (8) A simple funnel, which can be made from waste metal cans, speeds up pot filling if inserted in the polyethylene tube. Its lower opening should be just a little smaller than the diameter of the pots to be filled.

Scoop (9) Made of metal sheet, it is an efficient tool for filling and compacting soils in pots.

Pot cutting roll (10) Polyethylene pots are often sold in the form of rolls of endless tubes. A simple cutting gauge which permits rapid cutting of the tube into pots of standard length (15, 20 or 30 cm) can be made locally. A debarked piece of roundwood into which a slot has been carved is inserted into a support. The tube is then wound on to the piece of roundwood and cut (15-20 pieces at a time). To obtain pots of 20 cm length, the piece of roundwood has to have a diameter of 6.5 cm for 30 long pots of 8.5 cm.
A more comfortable device is obtained if the piece of roundwood is inserted in a frame and fitted with a handle.

TOOLS REQUIRED

(7) Sieve

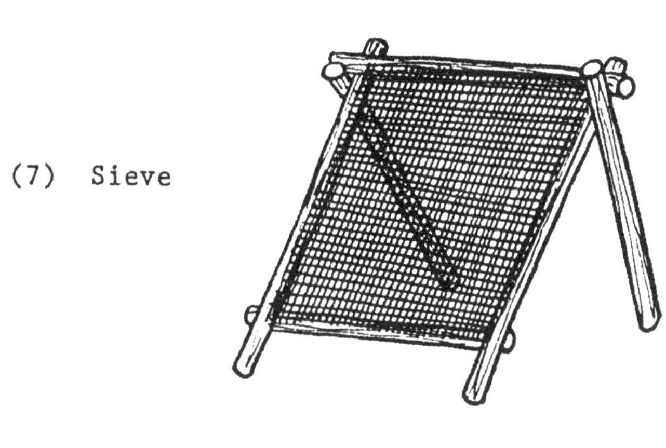

(8) Funnel

(9) Scoop

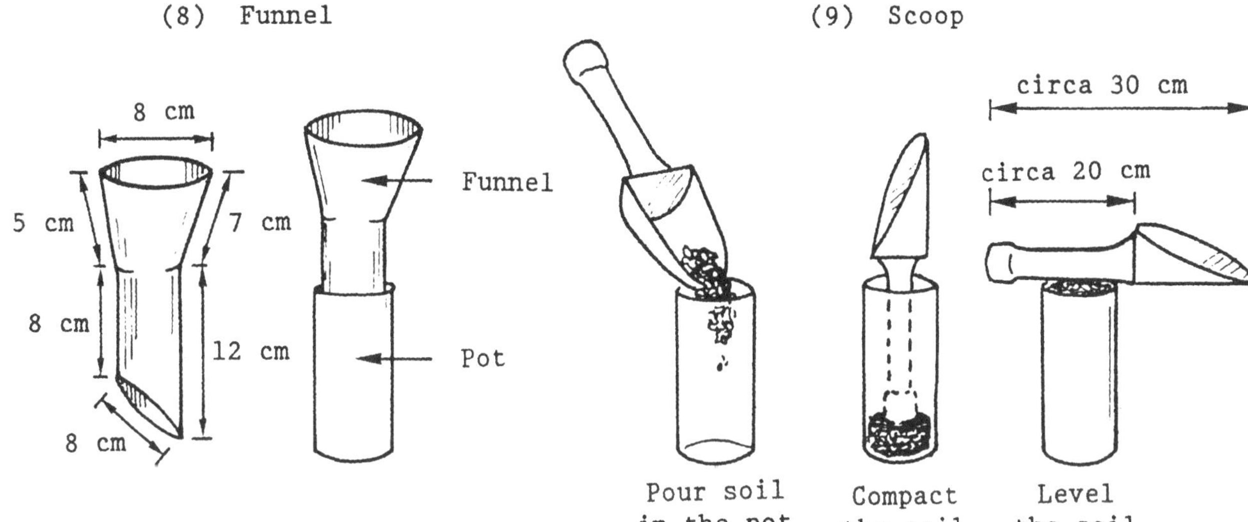

(10) Pot cutting roll

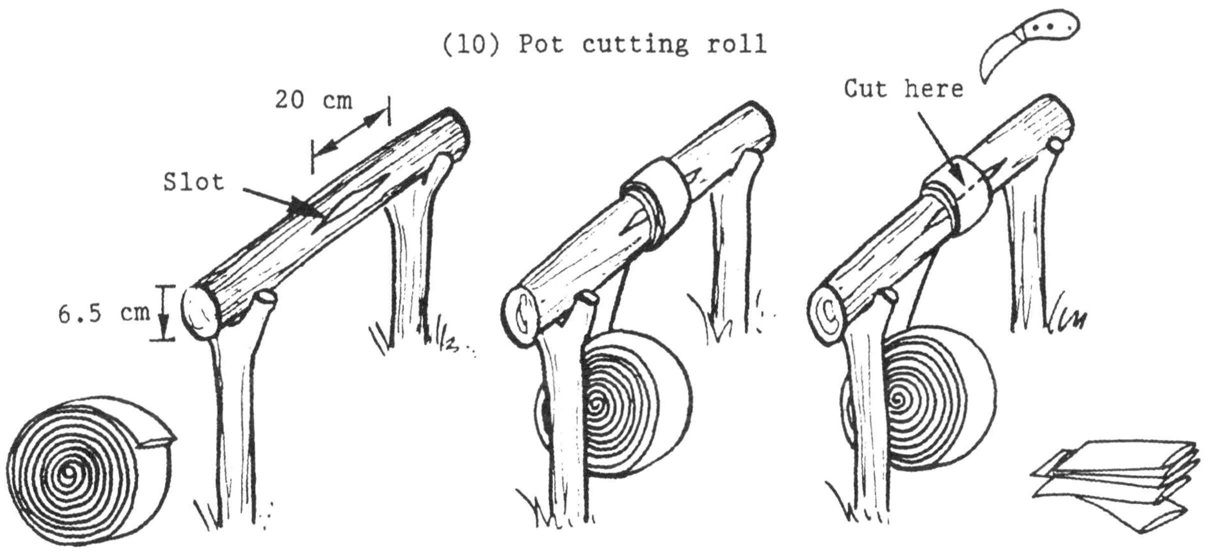

1. SETTING UP A TREE NURSERY

1.4 TOOLS REQUIRED

(d) FOR WATERING

Watering can (11) Watering cans can be made of metal or plastic. The latter has the advantage of being lighter. They should have a capacity of 10-12 l. It is important that they can be fitted with a finely-perforated sprinkler to avoid damage to young plants and denudation of roots from splashing water. The cans are easier to carry if there is a pair for each worker. A bucket yoke is a particularly useful device for carrying.

(e) FOR TRANSPORT WITHIN NURSERY

Wheelbarrow (12) This is most useful for the transport of all kinds of materials in the nursery (potting soil, filled pots, tools, seedlings ready for delivery, etc.). A sturdy model fitted with a metal bucket should be chosen.

TOOLS REQUIRED

(11) Watering cans

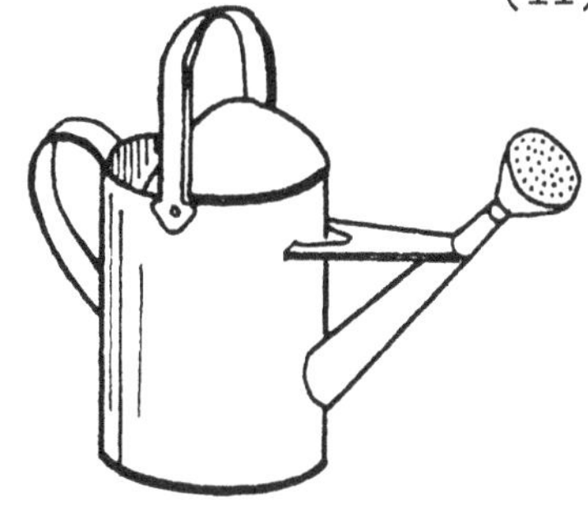

Ordinary watering can

Can converted into a watering can

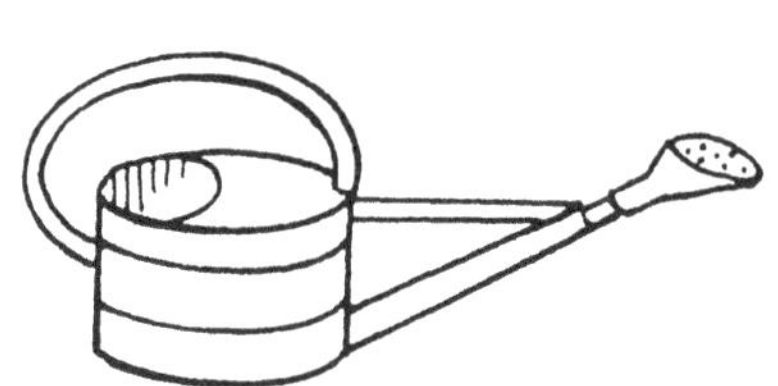

Small sprinkling can with horizontal nozzle to water delicate seedlings

Using a bucket yoke

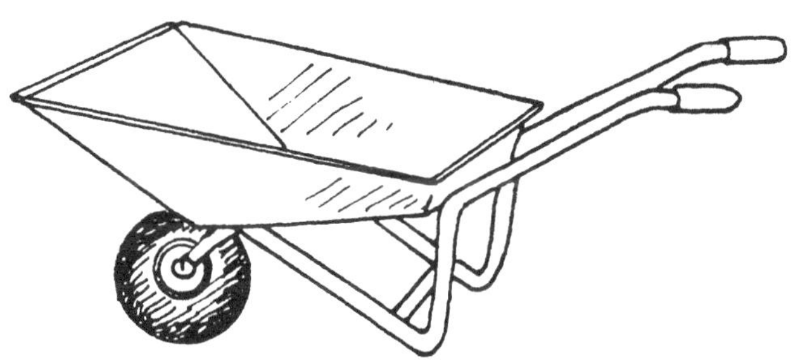

(12) Bucket wheelbarrow

1. SETTING UP A TREE NURSERY

1.4 TOOLS REQUIRED

(f) FOR TENDING OF SEEDLINGS

Pruning	(13)	Knives, shears, trowels or flexible steel wire can be used to prune the roots that grow out of the pot into the ground of the potbed.
Machete	(14)	Long knife which can be used for many purposes, such as cutting fence posts, removing woody weeds, trimming living fences, chopping left-over seedlings for composting, etc.
Tools for weeding	(15)	Simple tools like a pointed stick or a piece of strong wire hammered flat at one end and bent to make a handle at the other are useful for weeding seedbeds and potted stock.
Tools for pricking out	(16)	A small shovel, a flat piece of wood or simply a spoon are useful to lift germinated seedlings for transfer into pots without damaging their roots.

TOOLS REQUIRED

(13) Tools for root pruning

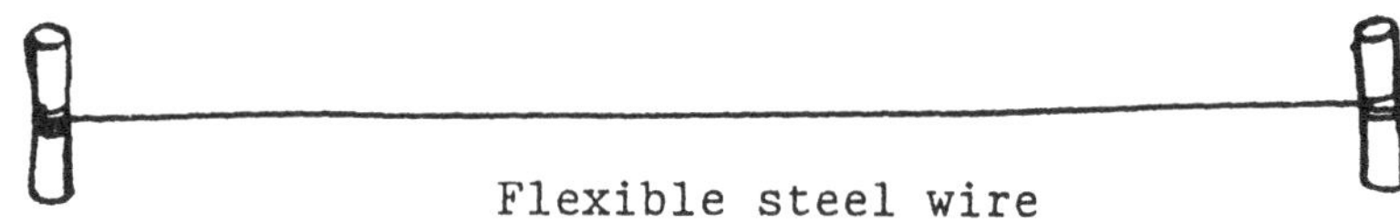

Flexible steel wire

Pruning knife

Trowel

Pruning shear

(14) Machete

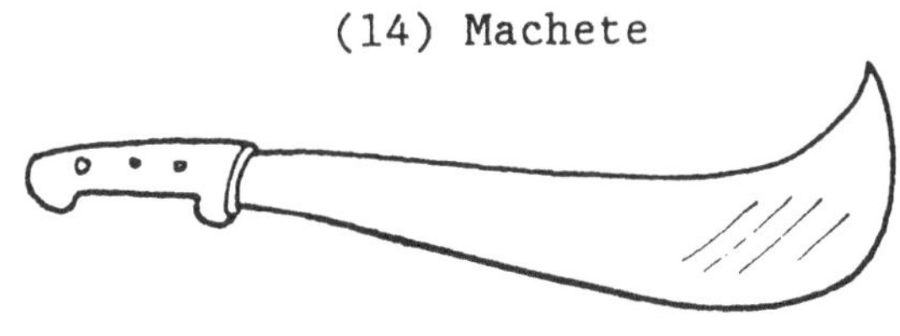

(15) Tools for weeding

Pointed wooden stick

Tool for weeding and cultivating soil surface in potted stock (made of strong wire)

(16) Tools for pricking out

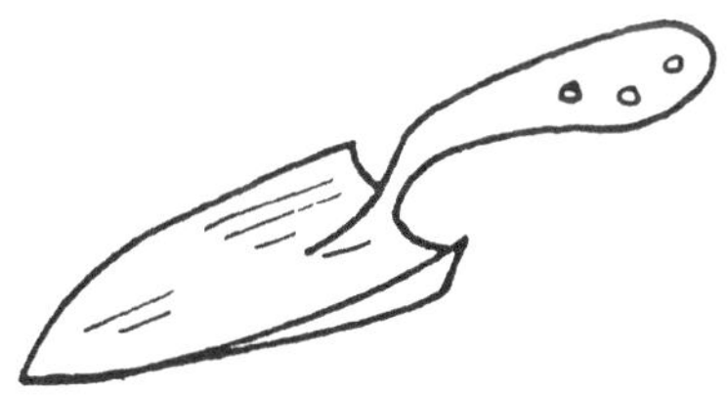

Small shovel

Flat piece of wood

Spoon

1. SETTING UP A TREE NURSERY

1.4 TOOLS REQUIRED

MOST COMMON ERRORS

1. Tool handles are often of inconvenient size, poor shape and badly fixed.

2. Cutting edges are not regularly sharpened.

2. SEED

2. SEED

2.1 WHERE TO GET SEED

Most planting stock is raised from seed. Successful raising and growing of trees depends on:

- the right kind of seed (provenance)
- good quality
- sufficient amount
- available at the right time.

Seed can be obtained from distributors or collected locally. In many countries, forest seed centres have been established by the Forest Service, universities or research institutes. Sometimes seed is sold by commercial firms or is available from other nurseries or projects. Any seed distributor must provide information on the species, when the seed was collected, where and from what type of parent trees.

The planting programme has to make sure the right species have been chosen. For the same species, there may be differences between trees growing at different altitudes and on different soils. Seed from the same species which originates from a specific place and is adapted to specific conditions, or has specific growth habits, is treated as a specific provenance. The place where the seed is obtained should be as similar as possible to the place where the seedlings will be planted.

If the seed needed is not available from distributors, it may be possible to collect it locally. To do this, the right place and good parent trees have to be selected. One also has to know when the trees will bear fruit.

It is advisable to establish a calendar of seed collection for the most important species in the project area.

WHERE TO GET SEED

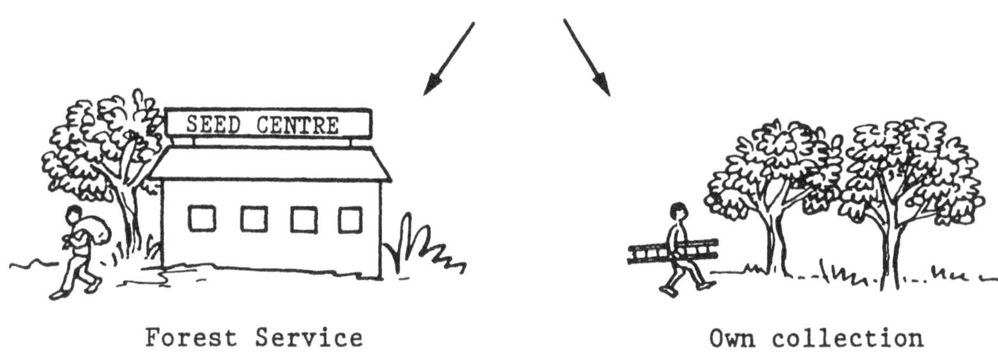

Forest Service

Own collection

PROVENANCES

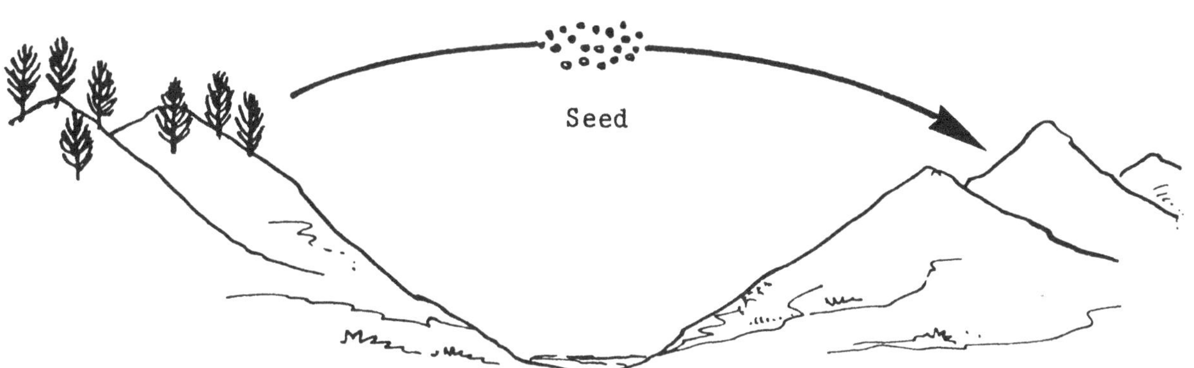

e.g. High altitude seed to high altitude planting site

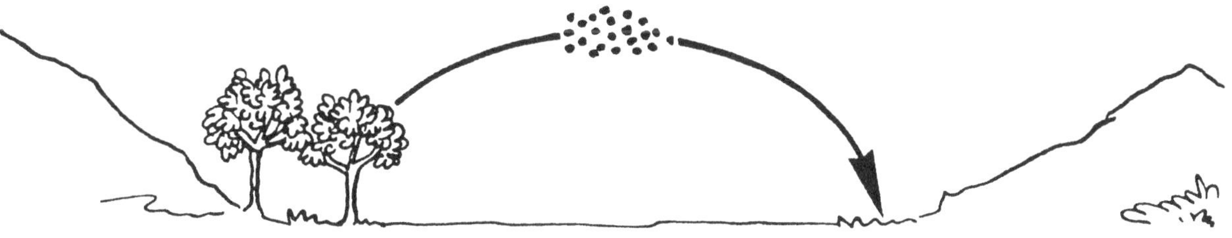

Seed from lowlands to lowland planting site

SEED TAG

Species:
Date of collection:
Origin of seed:
Purity:
Germination rate:
Seed lot:

2. SEED

2.2 SEED QUALITY

Good parent trees are important because the seedlings will be similar to the trees from which seed has been collected. Big, straight and vigorously growing trees will normally give straight and vigorously growing seedlings. Which characteristics to look for in parent trees will depend on the purpose of the plantation, however. Straight plants are important for construction wood or timber. For fuelwood production, fast growth and ready coppicing matter more. Ample leaf production and good regeneration after browsing or pruning are essential in fodder trees and shrubs.

Small seeds have to be collected from the trees by picking or cutting branches. For species with heavy fruits which will fall close to the tree it is also possible to collect from the ground, preferably by spreading large pieces of cloth under the tree.

In the latter case, the first and last fruits which fall should not be collected because they are often dull or damaged.

SEED CERTIFICATE   Name of institution ____________________

Species (botanical name) ________________________

(common name) ________________________

Date of collection __________   Batch no. __________

Country of origin __________   Area of origin ________

Detailed location   Latitude
Longitude
Elevation

Site: Soil type __________   Drainage ________   Slope __________

Stand: Age ______________   Mean height ______________

Diameter range __________   Stem form ______________

Remarks:

Certified by: ______________

THE RIGHT PARENT TREE

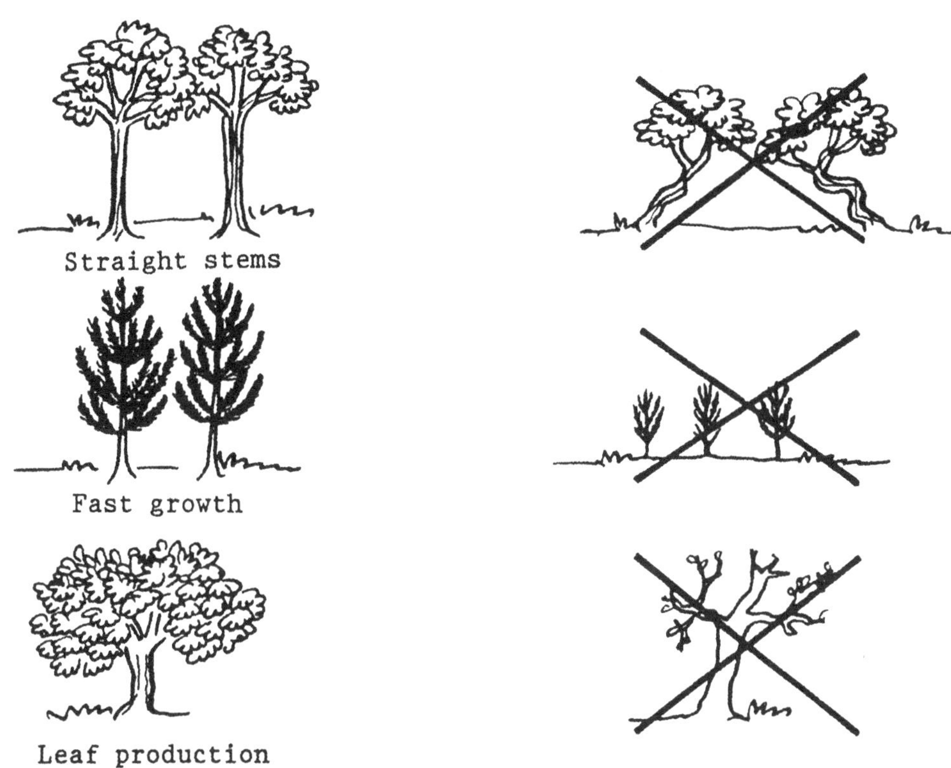

Straight stems

Fast growth

Leaf production
Dense foliage (fodder trees)

COLLECTION OF SEED

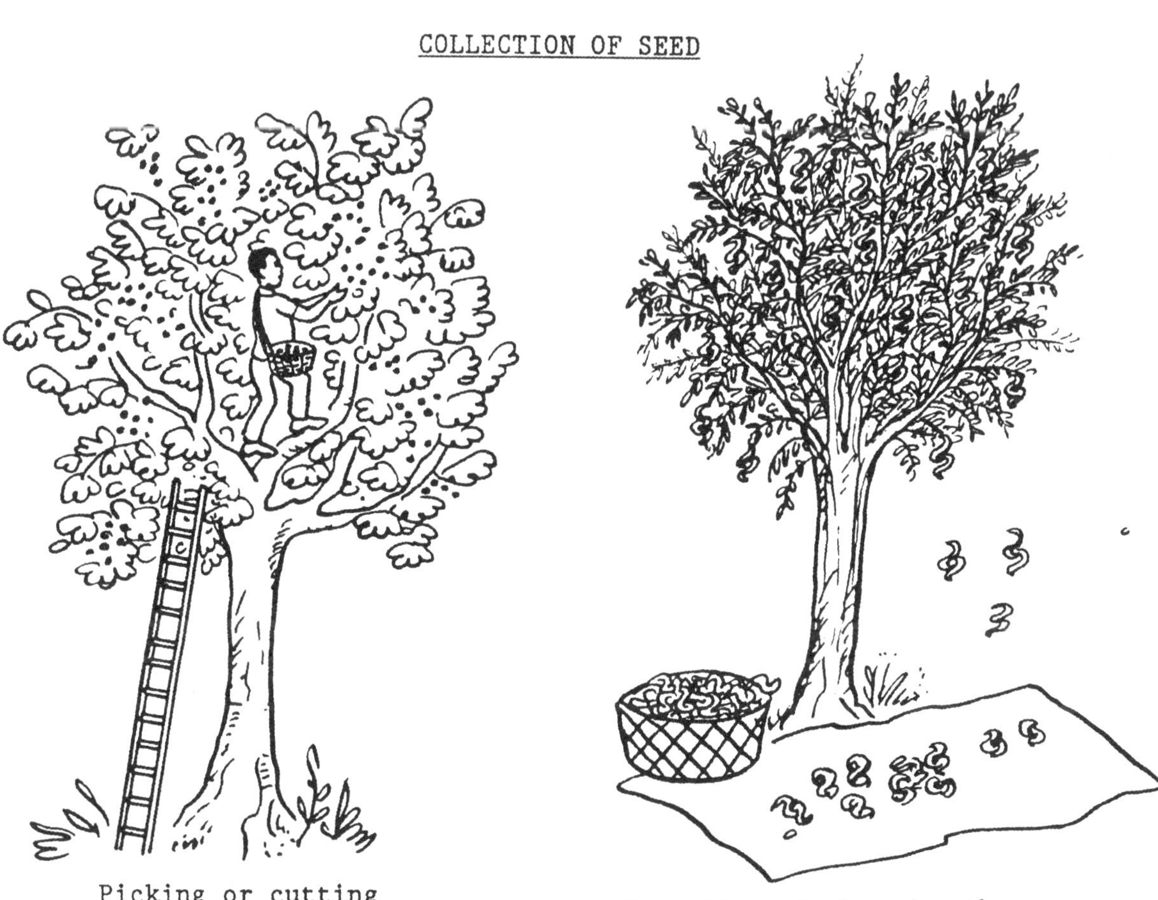

Picking or cutting

Spreading cloth [illegible]

2. SEED

2.3 HOW MUCH SEED WILL YOU NEED?

The amount of seed required depends on the number of seedlings needed for planting and the capacity of the seed to germinate and grow into a healthy seedling. The latter is known as the "germination rate" or "plant per cent" and is usually indicated on seed tags when seeds are obtained from distributors. Otherwise, the Forest Service or other projects will have rough figures from their experience. One can then calculate the amount of seed in kg as:

$$\text{Amount} = 125 \frac{N}{pW} + E$$

where N = no. of seedlings required for planting

p = germination rate or plant per cent

W = no. of seeds per kg

125 = a factor that adds a 25 per cent reserve

E = an extra quantity if some of the seeds are no good.

Example of calculation of seed requirement

$$\text{Amount (in kg)} = 125 \times \frac{N}{pW} + E$$

- for species A planting target 20 ha
  planting density 2,500 seedlings/ha

- number of plants required (N) = 20 x 2,500
  = 50,000 seedlings

- germination rate (p) = 20%

- number of seeds per kg (W) = 70,000

- 10% will get damaged during storage or development (E) = 10%

- amount (in kg) including 25% reserve =

$$125 \times \frac{50{,}000}{20 \times 70{,}000} = 125 \times \frac{50{,}000}{1{,}400{,}000} =$$

$$125 \times \frac{15}{1{,}400} = 125 \times 0.0107 = 4.464 \text{ kg}$$

$$4.464 \text{ kg} + 10\% = 4.464 + 0.446 = 4.910 \text{ kg}$$

say, approximately 5 kg.

EXERCISE: Work out the quantity of seed needed to plant 10 ha of a mixture 50% of species B and 50% species C at each 1,100 trees/ha. The germination rate is 35% and there are 300,000 seeds/kg of species B and the germination rate is 40% and there are 70,000 seeds/kg of species C. A reserve of 20% will be needed for both.

N =
p =
W =
E =

Result:

2. SEED

2.4 EXTRACTION AND STORAGE

Extraction

The seed of most species can be extracted by sun-drying. The collected fruits are spread in thin layers on canvas, trays or the like, exposed to the sun, and stirred regularly until cones, pods or capsules open and release the seed. For species with fleshy fruit, the pulp has to be removed first. Chaff and seed can be separated by winnowing or by submersion in water. Good seed will usually sink to the bottom, while dull seeds, chaff and other impurities float. After separation, good seeds are thoroughly dried in the sun. Moist seed cannot be stored!

Storage

For most species, cool and dry storage in a dark place is best. Some species can be stored for several years without much loss of viability but some will lose the potential to germinate within a few months. Seed of leguminous species (those having pods like beans) can be stored for many years even at normal temperature. For any seed, storage temperature should be kept as constant as possible even if cool storage cannot be provided.

Seed for storage must be dry. The store (if they are not kept in a refrigerator) must be ventilated. Some seeds are eaten by rodents or insects and have to be protected. Seeds sensitive to moisture and those stored for a longer time are packed in air-tight bottles or tins. Other seed can be kept in wooden boxes or sacks made of cotton. No plastic boxes or sacks can be used unless packing is air-tight because the lack of ventilation makes seeds damp and leads to moulds.

EXTRACTION AND STORAGE

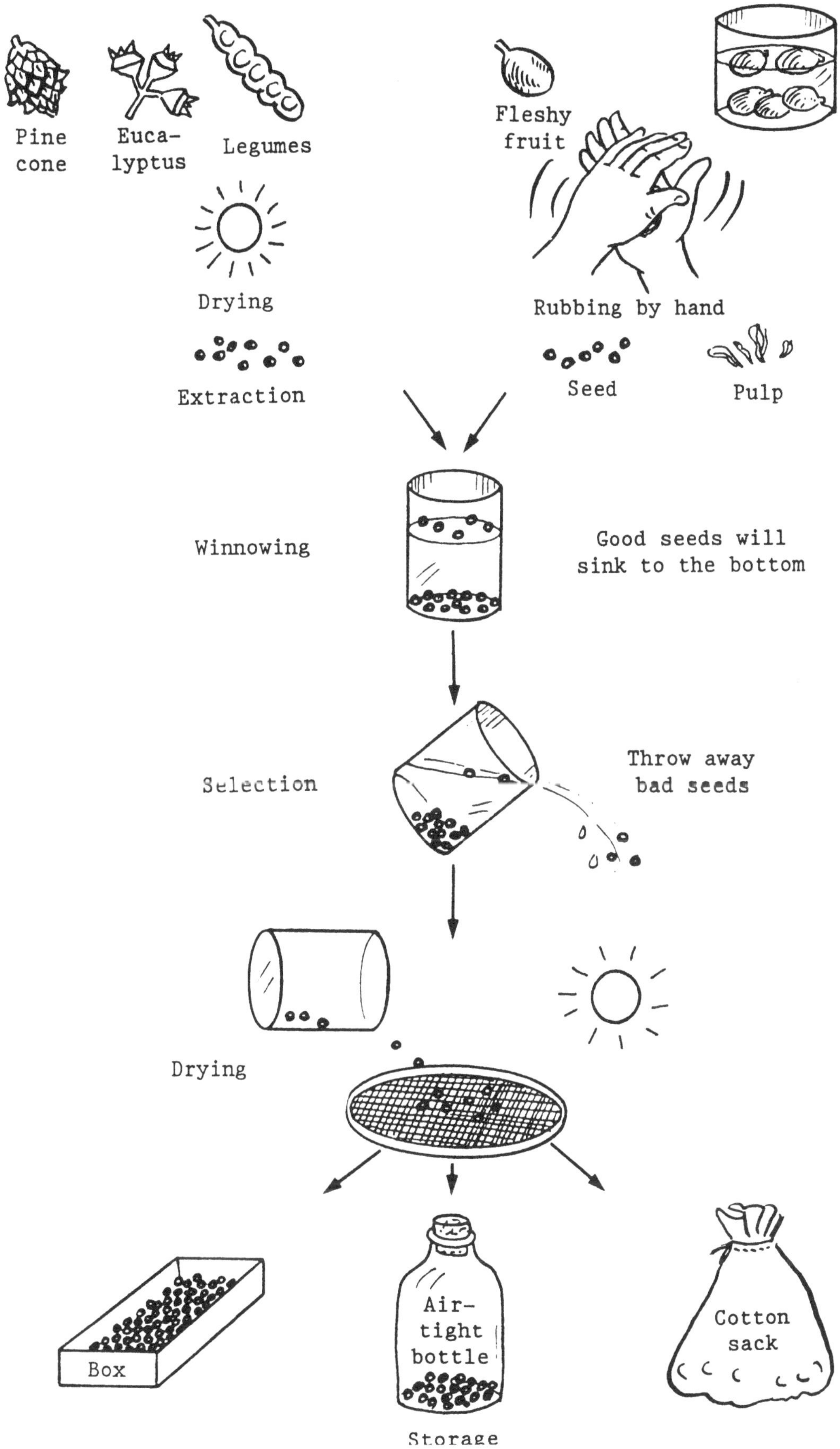

Pine cone

Euca-lyptus

Legumes

Fleshy fruit

Drying

Rubbing by hand

Extraction

Seed

Pulp

Winnowing

Good seeds will sink to the bottom

Selection

Throw away bad seeds

Drying

Box

Air-tight bottle

Cotton sack

Storage

2. SEED

2.4 EXTRACTION AND STORAGE

MOST COMMON ERRORS

1. The collection of seeds from small, crooked, unhealthy trees because this is easier. The seedlings are likely to have the same poor growth and liability to disease as the parent plants.

2. Missing the right time for collection and then producing whatever species of seed is available rather than species selected for the sites and the purpose of the plantation.

3. GROWING SEEDLINGS IN CONTAINERS

3. GROWING SEEDLINGS IN CONTAINERS

Growing seedlings in containers is the standard method in most projects. It requires the following eight operations:

OPERATION 1: OBTAIN POTTING SOIL

OPERATION 2: PREPARE SUITABLE MIXTURE

OPERATION 3: FILL POTS

OPERATION 4: PLACE POTS IN BEDS

OPERATION 5: SOWING

OPERATION 6: PRICKING OUT

OPERATION 7: CARE AND TENDING

OPERATION 8: PREPARATION FOR PLANTING OUT

GROWING SEEDLINGS IN CONTAINERS

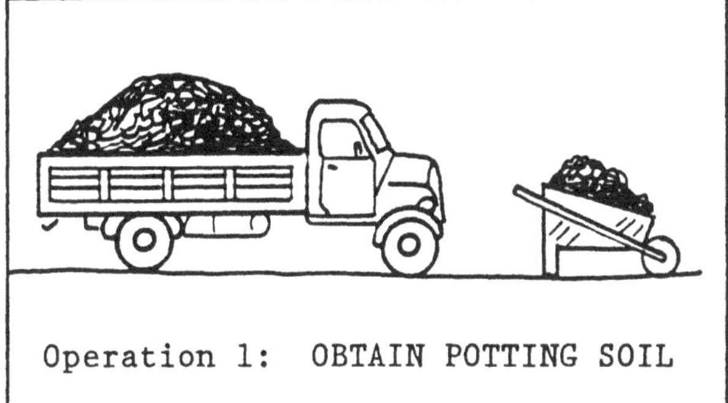

Operation 1: OBTAIN POTTING SOIL

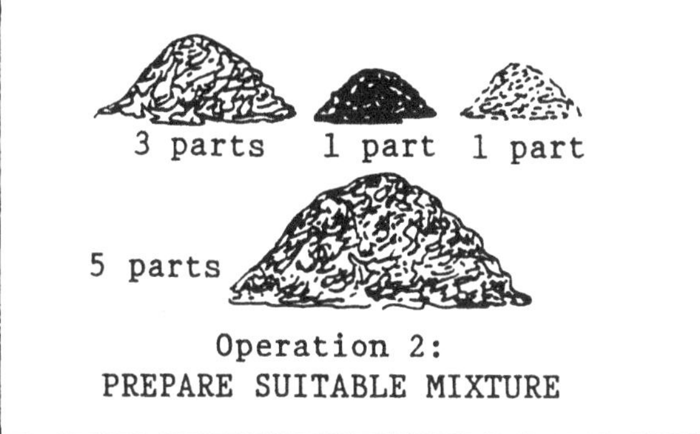

Operation 2:
PREPARE SUITABLE MIXTURE

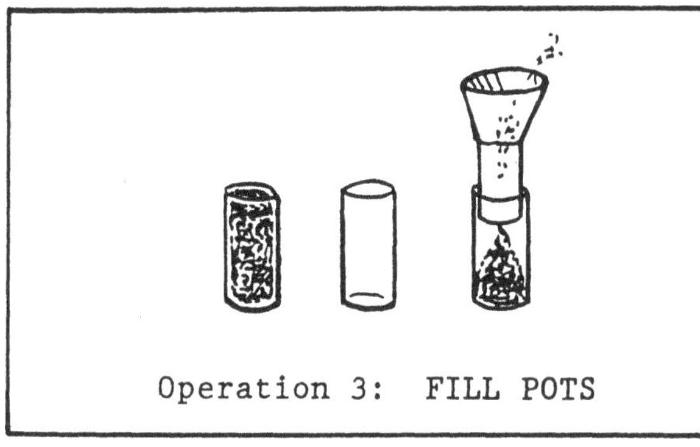

Operation 3: FILL POTS

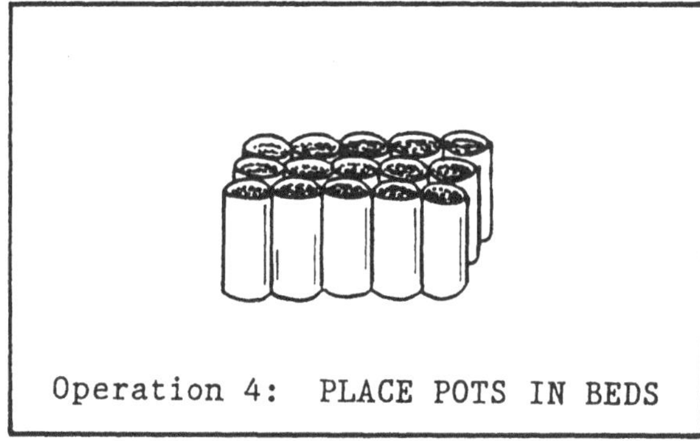

Operation 4: PLACE POTS IN BEDS

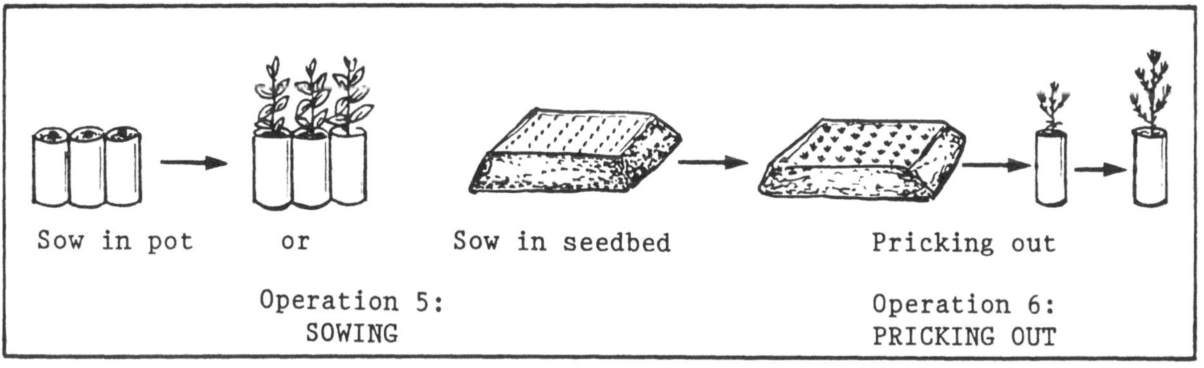

Operation 5:
SOWING

Operation 6:
PRICKING OUT

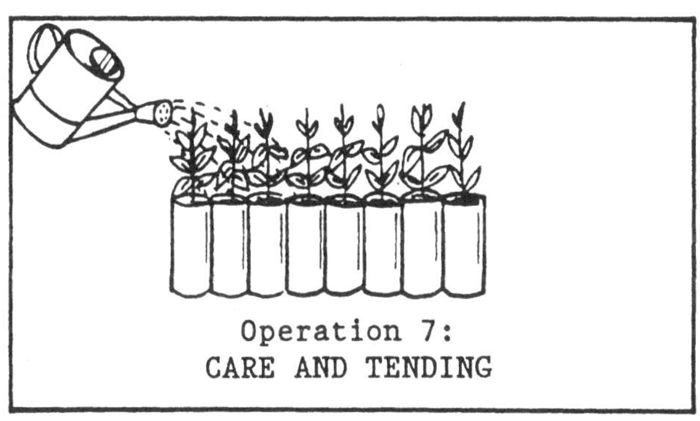

Operation 7:
CARE AND TENDING

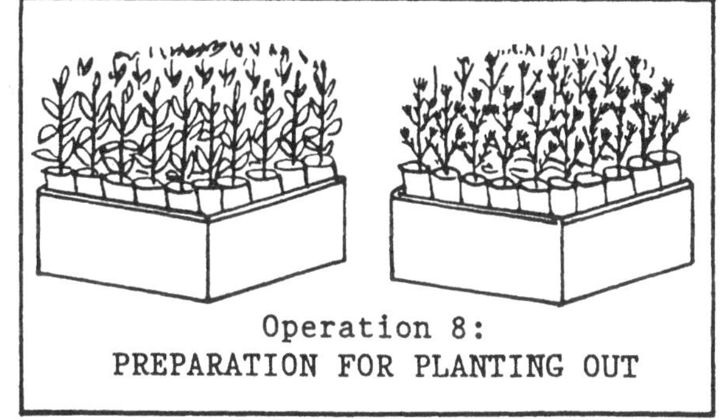

Operation 8:
PREPARATION FOR PLANTING OUT

3. GROWING SEEDLINGS IN CONTAINERS

OPERATION 1: OBTAIN POTTING SOIL

The soil in the pots has to facilitate germination and root development and to supply the seedling with water and nutrients. It should therefore be light and rich in nutrients. On the other hand, the soil should not fall out of the pot during handling and transport or crumble too easily when planted. Soil with a lot of weed seed in it should be avoided. Soil from underneath leguminous trees such as Acacia is particularly rich in nutrients.

Suitable soil has to be brought with a cart or truck, or possibly a wheelbarrow if it is not available at the nursery site. Usually different types of soil have to be mixed to obtain potting soil. Successful mixtures can normally be made from local material and little transport should be required.

The quantity required depends on the number of seedlings to be produced as well as on the size of the containers. A table for calculation, as well as an example of its use, is provided in Appendix 3.

SOURCES OF SOIL

Humus-rich soil

Under trees

In forest

Agricultural soil

Fields

Gardens, orchards, fallow land

Sand

River bed

Compost, manure, peat

3. GROWING SEEDLINGS IN CONTAINERS

OPERATION 2: PREPARE SUITABLE MIXTURE

The mixture used will normally contain:

- humus-rich soil as found under trees or in forests; and
- ordinary agricultural soil as found in crop fields, gardens or fallow land.

If soils contain too much clay so that they are heavy and crack when dry, sand from riverbeds or dunes is added.

If the soils are not rich enough in humus, well rotten animal manure, compost or peat are added. Compost making is described in Appendix 1.

Some tree species need fungi and bacteria in the soil for good development. If these do not exist in the soil because the tree species do not grow where the soil has been taken from, then they should be added. Many pine species need fungi called mycorrhiza. Appendix 2 gives more detail on these.

Since the mixture depends very much on the quality of the soil available, no standard recipe can be given. If in doubt, try a mixture of:

3 parts agricultural soil
1 part forest soil/manure
1 part sand.

PREPARE SUITABLE MIXTURE

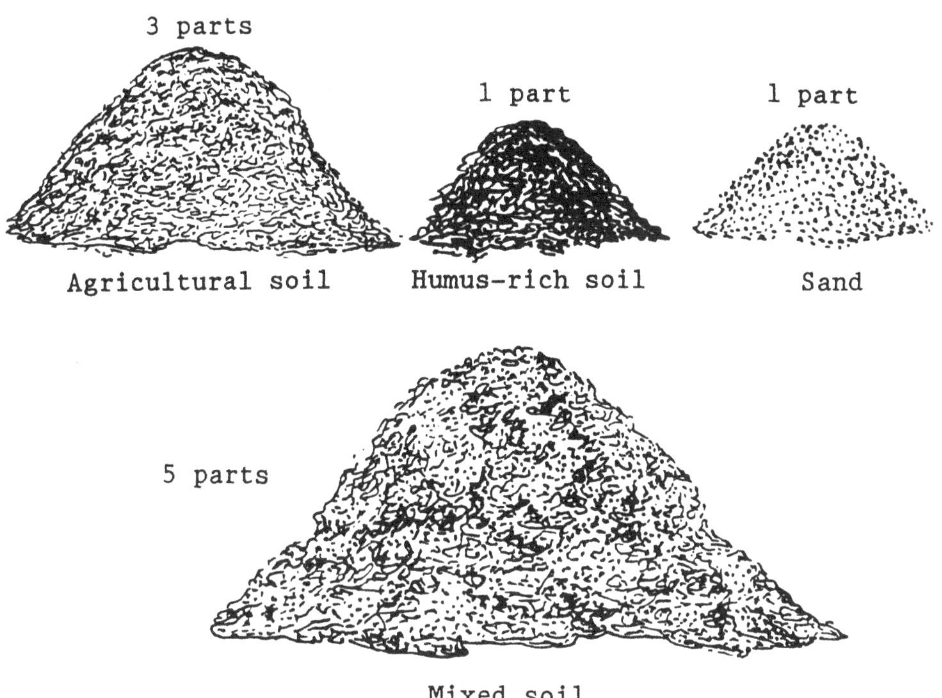

WHERE TO GET FUNGI AND BACTERIA

Fungi (mycorrhiza) in soil

3. GROWING SEEDLINGS IN CONTAINERS

OPERATION 3: FILL POTS; OPERATION 4: PLACE POTS IN BEDS

OPERATION 3: FILL POTS

When the soil, sand, manure or compost have been obtained, they are sieved (see page 24 for sieve) and mixed thoroughly. The mixture is then moistened to become humid but not wet. The cutting of pots is explained on page 24.

The pots are filled by hand or with the help of a funnel (see page 24). When filling, the lower third of the container should be compacted rather firmly to make sure the soil does not fall out again easily. The upper two-thirds are compacted only gently so that roots can develop more easily. However, air-pockets should not be left inside the container during filling, since these hamper root development. The workers should be comfortably seated whilst performing this operation to avoid fatigue.

OPERATION 4: PLACE POTS IN BEDS

Finally, the pots are placed in the beds in an upright position. They should not be squeezed but maintain their round shape and space should be left for rain and excess water to drain off easily.

MOST COMMON ERRORS

The soil is too heavy (too much clay) so drainage is poor, shoots have difficulty in emerging and roots do not develop well. Germination rate will be low and germination will take more time.

FILLING POTS

Correct

False:
too compact

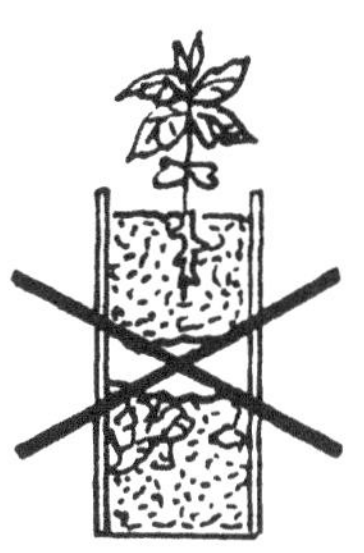

False:
air pockets

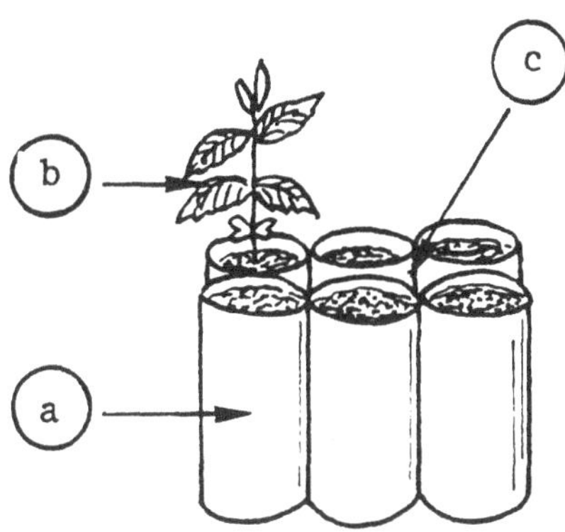

Correct:
(a) Pots are not deformed
(b) Seedling in centre of pot
(c) Pots placed in straight rows leaving space for draining

False:
(d) Pots are deformed
(e) Seedling not placed in centre
(f) Pots placed too densely, insufficient space for drainage

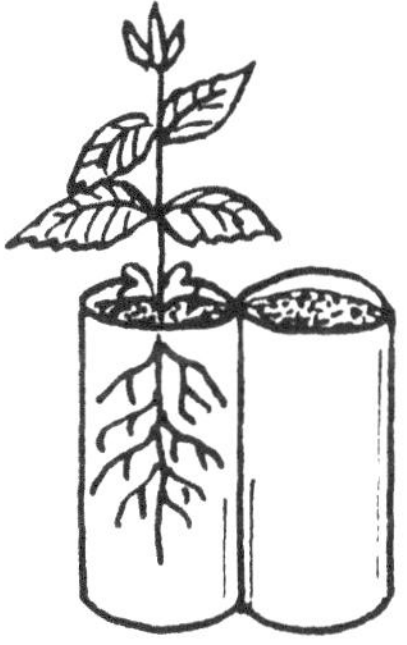

Correct:
pot upright -
roots will develop well

False:
pots inclined -
roots distorted

3. GROWING SEEDLINGS IN CONTAINERS

OPERATION 5: SOWING

Sowing consists of four steps:

STEP 1: CHOICE OF SOWING METHOD: IN POT OR SEEDBED?

For species which germinate well and more or less at the same time, direct sowing into pots is the most convenient. Examples are leguminous species (*Acacia*, etc.) and *Eucalyptus*.

For species which germinate with difficulty, which need special care or of which seed is rare and expensive, sowing into a seedbed is preferable. The layout of a seedbed and its composition of layers - gravel, humus-rich soil and a mixture of sand and soil - are shown on page 11.

MOST COMMON ERRORS

1. The seedbed surface is not well levelled but irregular or even shaped like a mound. Watering will be uneven and cause erosion as well as washing away the seed.

2. Seedbed soil not changed for many years. This favours the development of diseases.

STEP 2: PROCUREMENT OF SEED IN THE REQUIRED QUALITY AND QUANTITY

Following the explanations provided about seed in Chapter 2, the required amount of seed is prepared.

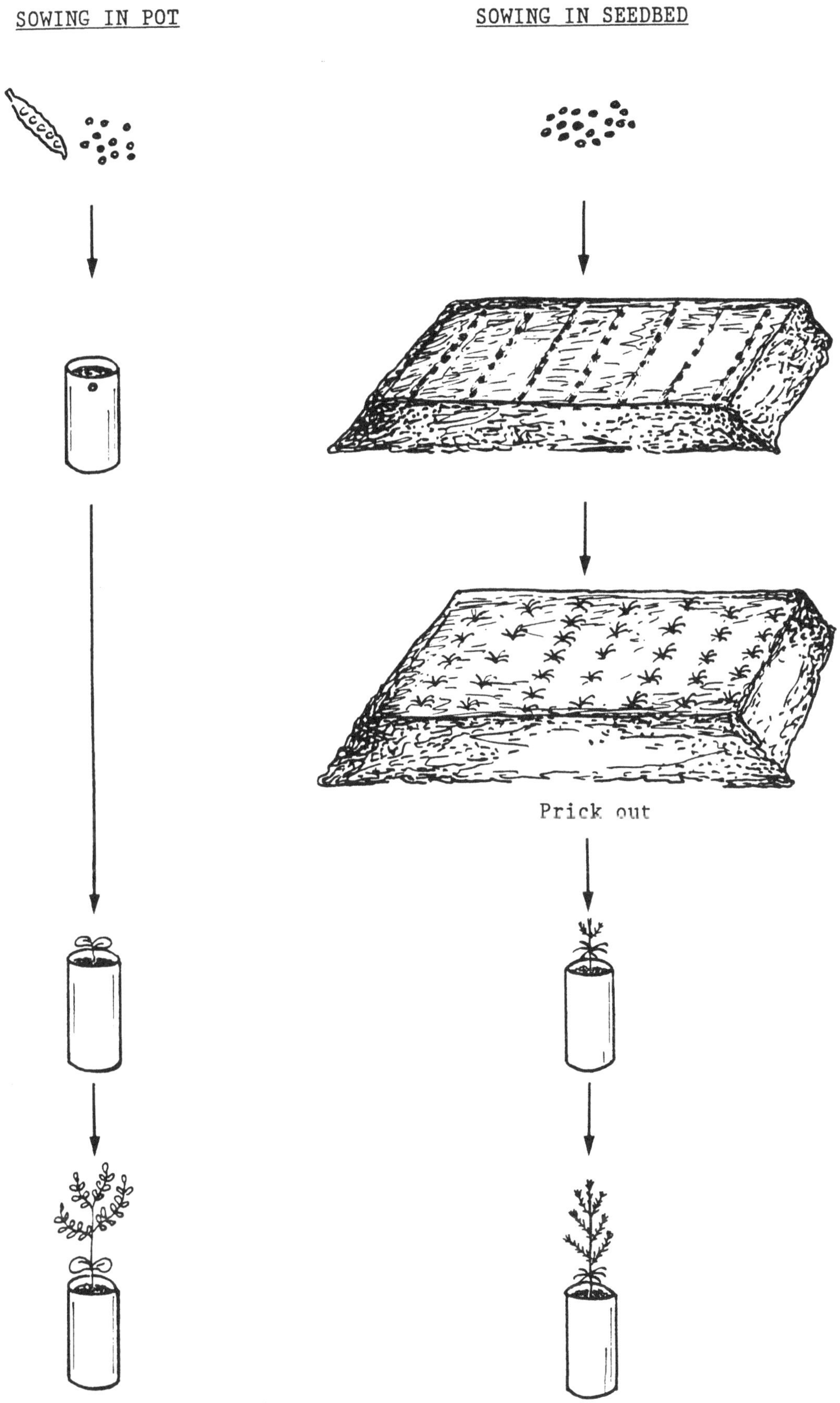

SOWING IN POT

SOWING IN SEEDBED

Prick out

3. GROWING SEEDLINGS IN CONTAINERS

OPERATION 5: SOWING

STEP 3: PRE-TREATMENT OF SEED

Some species do not germinate or only after a long time if they are not pre-treated. The simplest way is to soak the seed in cold water for two days. Leguminous species like Acacia, Cassia, Prosopis have to be treated with hot water. For this:

- bring water to the boil (about 20 l for 5 kg of seed)
- take the water off the fire
- immerse the seed
- allow to cool overnight
- rinse with clean water.

For other pre-treatments that may be required for local species, ask for advice from Forest Services or seed centres.

STEP 4: SOWING

For sowing in pots, place a number of seeds into each pot depending on the germination rate.

For sowing in seedbeds, the seed is best put in rills running across the bed because weeding is easier.

In both cases, seed has to be placed at the right depth. As a rule, seeds should be one or two times their diameter under the surface. Very small seed, like that of Eucalyptus, can be mixed with sand before sowing to make even distribution easier.

The seeds are covered with sand or soil which is gently pressed.

The seedbeds or pots are watered with cans with fine nozzles.

Finally, they are covered with shading material.

HOT WATER TREATMENT

SOWING DEPTH

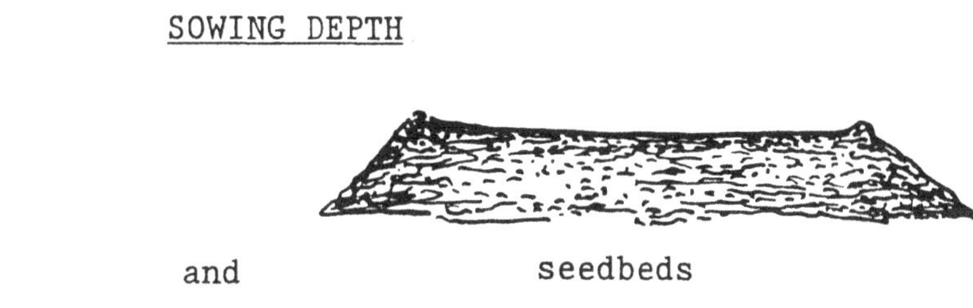

Eucalyptus

Pinus

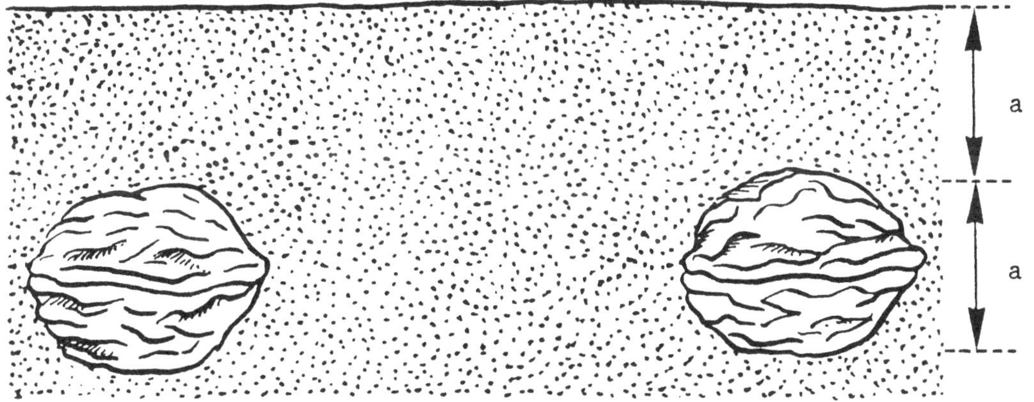

Big seeds, e.g. Juglans

3. GROWING SEEDLINGS IN CONTAINERS

OPERATION 5: SOWING

MOST COMMON ERRORS IN SOWING

1. Sowing is too dense or too thin. Germination rates will be low and development of diseases is favoured if sowing is too dense. Seedling production is more costly if sowing is too thin.

2. Seed placed too deep. Germination rate will be low and germination will take more time.

3. The time of sowing is not chosen in accordance with the time needed to grow a tree of a size suitable for planting out. This is one of the most common and serious errors because it cannot be repaired. The seedlings will either be too small or too big. Chapter 7 explains how this can be avoided.

4. Large quantities of seed of the same species are sown at the same time. This leads to a labour peak for pricking out and all seedlings reach planting size at the same time when in fact planting takes place over several weeks. The seed should therefore be sown in several batches over a few weeks.

SOWING - MOST COMMON ERRORS

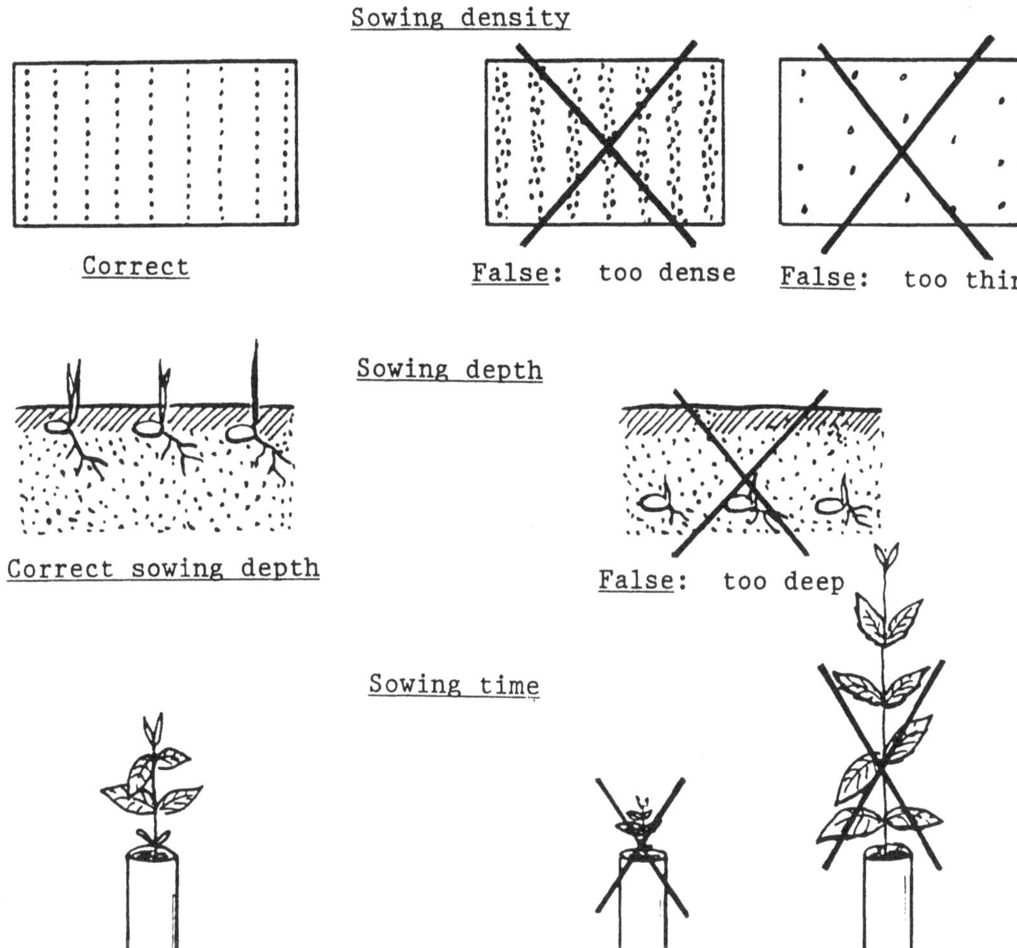

Correct: development as required for planting out

False: too late sowing

False: too early sowing

Spacing of sowing

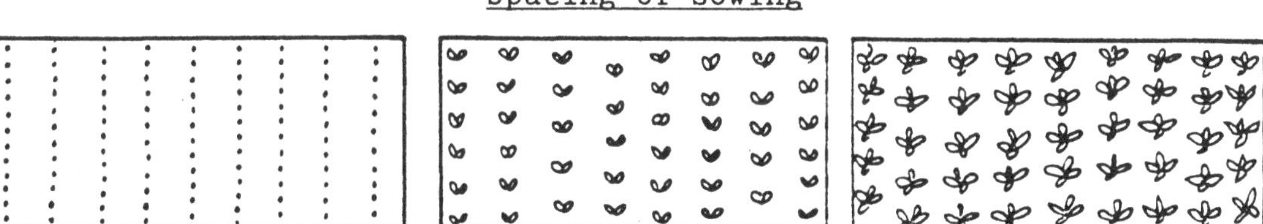

Correct: Seed is sown in several batches over a few weeks

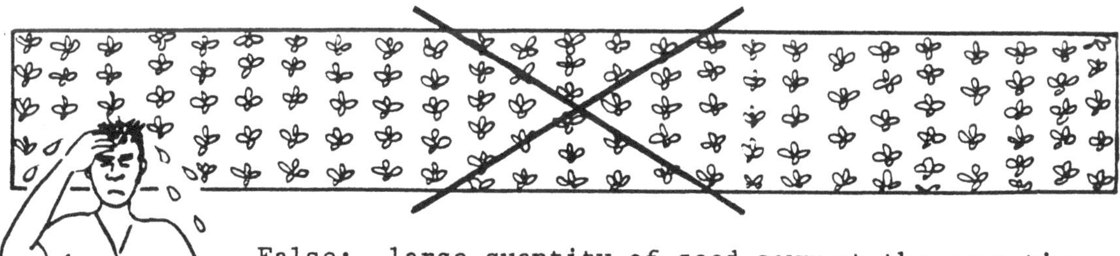

False: large quantity of seed sown at the same time causes problems!

3. GROWING SEEDLINGS IN CONTAINERS

OPERATION 6: PRICKING OUT

Sowing in seedbeds requires transfer of the germinated plants into pots after some time. This operation is known as pricking out. It is very important that the seedlings are pricked out when they have reached the right size. This is when the seedling has its first true leaves. These are the leaves which appear after the germination leaves. If transplanted earlier, the seedlings are too delicate. Later, their roots get too long or the seedlings stop growing because they are too close together.

Seedbed and pots should be watered thoroughly the day before pricking out to avoid damaging them. Seedlings are lifted with a little shovel or a flat piece of wood. Only healthy, well-developed seedlings are pricked-out. Unsatisfactory ones are thrown away. Always hold the seedling by its leaves, never by its stem, which is very delicate and will not recover if compressed. If roots are too long, prune them with a sharp knife. If only a tap root has to be cut, use your fingernails.

The seedlings dry out extremely quickly. The roots may die after only three minutes in the full sun or if exposed to dry wind. Therefore: work under shade; work in the late afternoon or when the sky is cloudy. Keep the seedlings under a moist cover (but not in water!). Make the dibble holes deep and wide enough to accommodate the roots. The roots should not be bent or point upwards. The hole is closed by gently pressing soil around the plant with the fingers. Water the plants and shade them.

MOST COMMON ERRORS

1. If carried out too late, roots will easily be damaged and work will be more difficult.

2. Roots are bent during pricking out either because it is carried out too late or poorly, e.g. into a hole that is too small. Many seedlings will not survive and others have root deformations.

3. Roots are too much exposed to sun or wind.

PRICKING OUT

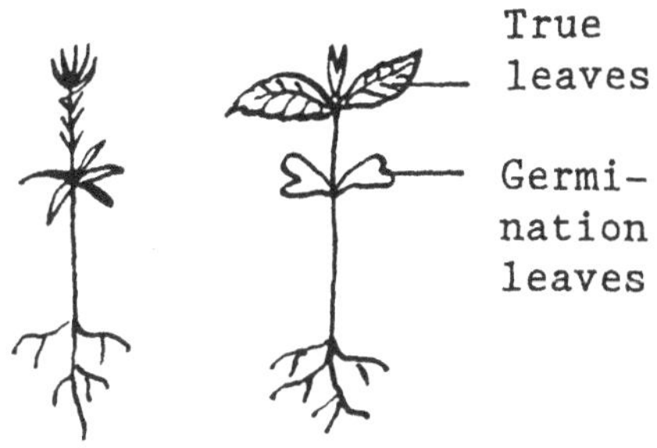

True leaves

Germi-
nation
leaves

Check size

Water

Lift seedling

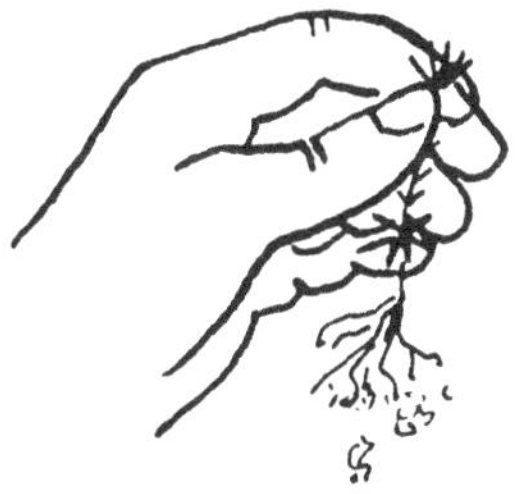

Hold the seedling by top leaves

Moist cover

Prune roots

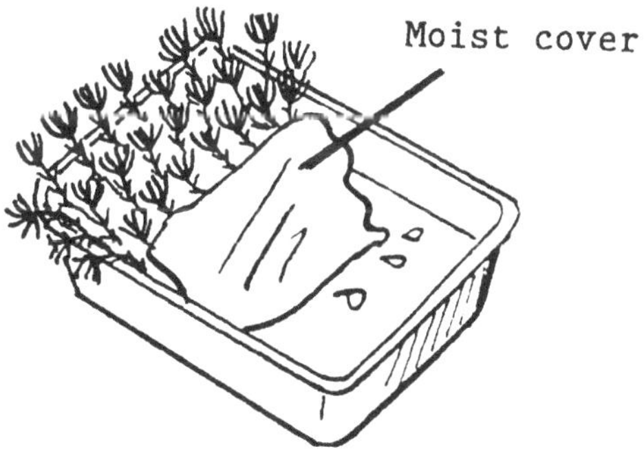

Cover plants

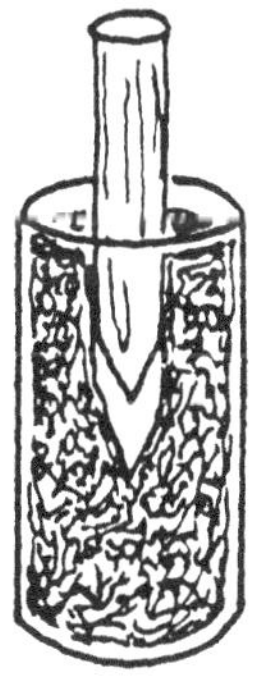

Dibble hole

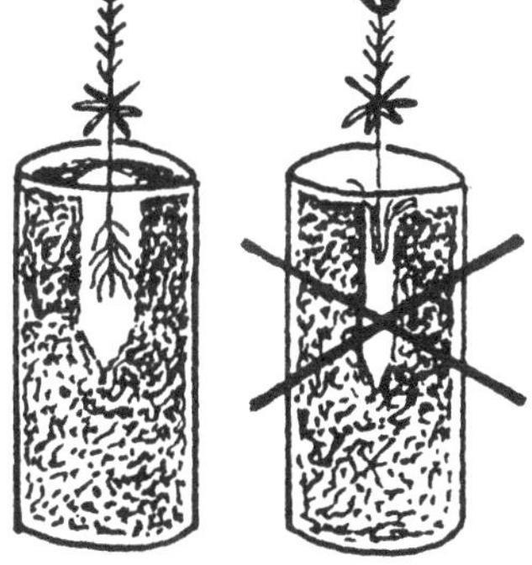

Insert plant -
make hole big enough!

Close hole

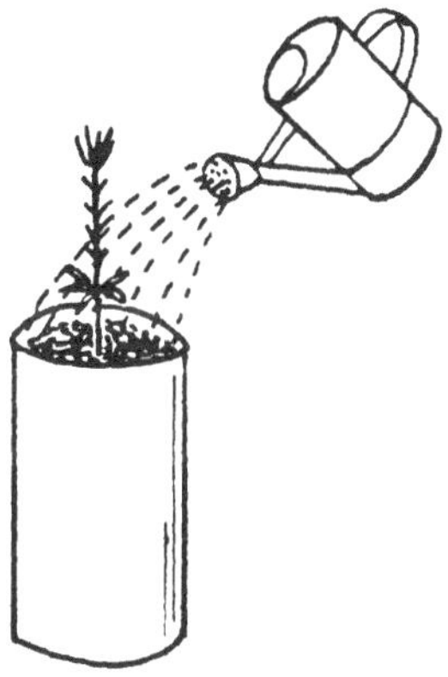

Water and shade

3. GROWING SEEDLINGS IN CONTAINERS

OPERATION 7: CARE AND TENDING

Care and tending of seedlings comprises five activities:

1. Watering
2. Weeding
3. Shading
4. Prevention and control of pests and diseases
5. Root (and possibly shoot) pruning.

Watering

A good supply of water is indispensable for the growth of the seedlings but too much water can be harmful. Therefore, observe the following:

- Water twice a day for the first weeks after germination because the roots do not reach deeply.
- Water only once a day but thoroughly later on. Make sure the water soaks the pots to the bottom.
- Before watering, check some pots in the bed. If the soil is still sufficiently moist, do not water.
- Water in the evening so less water will be lost through evaporation.
- Use watering cans or sprinklers with fine nozzles so as not to wash the soil out of the pots.
- Take care that all parts of a bed are watered evenly.
- If flood irrigation is used, make sure the water is drained from the beds after the pots have been soaked.

Too much water reduces root growth, favours fungi and makes the shoots of the seedlings long and soft. This should be avoided. Yellowish leaves and a cover of mosses on the tops of the pots are other signs of excess watering.

Weeding

Weeds take away light and water from the seedlings and increase the risk of fungi attacks. They have to be removed as they start to appear. When weeds become too tall, the roots of the seedlings are easily damaged during weeding and the work becomes more difficult, time-consuming and costly. The workers should have a little stool to sit on during manual weeding to allow for a convenient working position.

Weeds should never flower and set fruit in the nursery or the surroundings.

WATERING

First weeks after germination

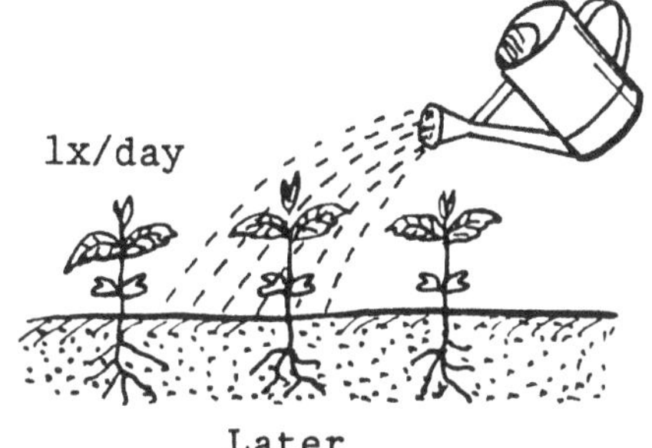

Later

Check some pots before watering

Water in the evening

Correct: fine nozzles

False: without nozzle

WEEDING

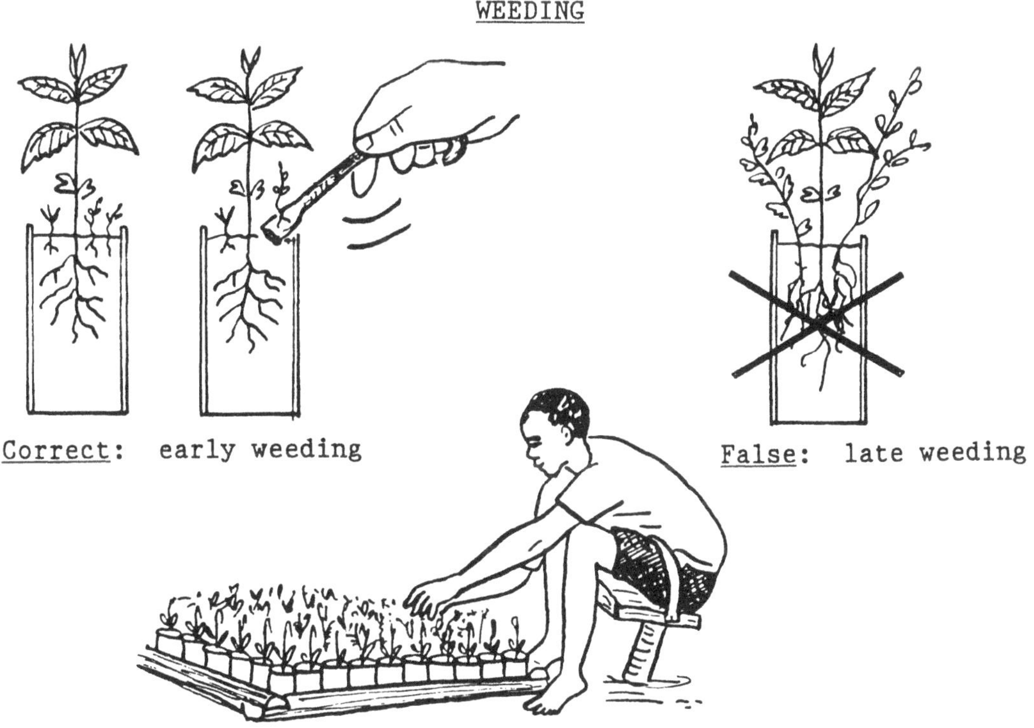

Correct: early weeding

False: late weeding

Little stool for weeding

3. GROWING SEEDLINGS IN CONTAINERS

OPERATION 7: CARE AND TENDING

Where sowing has been directly into pots, germinated seedlings are reduced to one per pot by removing the weaker and smaller ones.

Shading

In the early stages of their development - from sowing to some time after pricking out - seedlings are sensitive to full sunlight and high temperatures. During this phase, they must be protected by shading mats.

When the seedlings are more resistant, shading is reduced gradually from all day to around mid-day, and later to none at all.

For the last months in the nursery, seedlings should be exposed to full sunlight. Shade needs vary with species as well but observation of the plants will show what is needed.

Shading mats can also be used at any stage to protect seedlings from heavy rains or hail which can otherwise damage seedlings badly.

Prevention and control of pests and diseases

Small organisms like fungi, bacteria and viruses or animals like nematodes and insects can cause damage to seedlings. A very common disease in nurseries is caused by fungi and called "damping-off". It can cause seed to rot before germination, roots to decay before the shoot appears or the shoot to become thin at its collar and collapse.

Damping-off can often be prevented by:

- changing soil in seedbeds after some years;
- immediately removing plants infected by fungi or attacked by pests and burning or burying them;
- avoiding excessive watering;
- allowing for good drainage (sand in seedbeds);
- ensuring good air circulation;
- removing weeds and lowering density of seedlings in beds;
- reducing nitrogen content (less manure, no chemical fertiliser).

SHADING

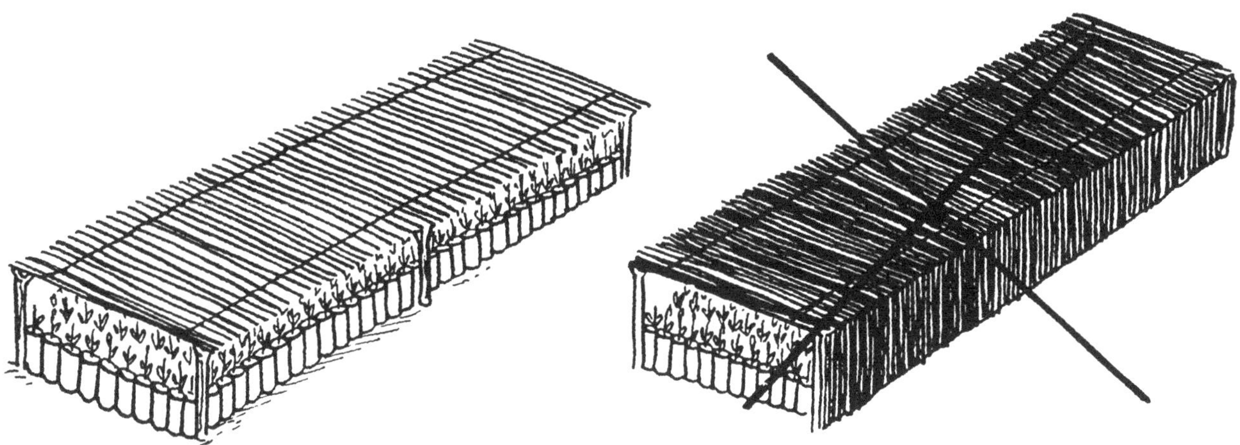

Correct: shading permitting good air circulation, removeable to allow bed to dry after excessive rainfall

False: excessive shading

DISEASES

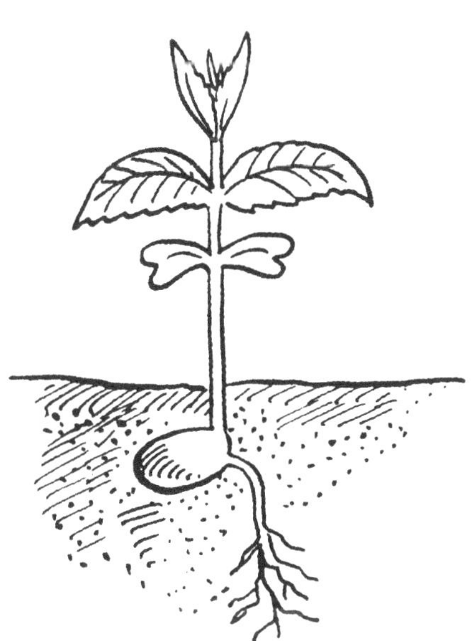

Correct: healthy plant

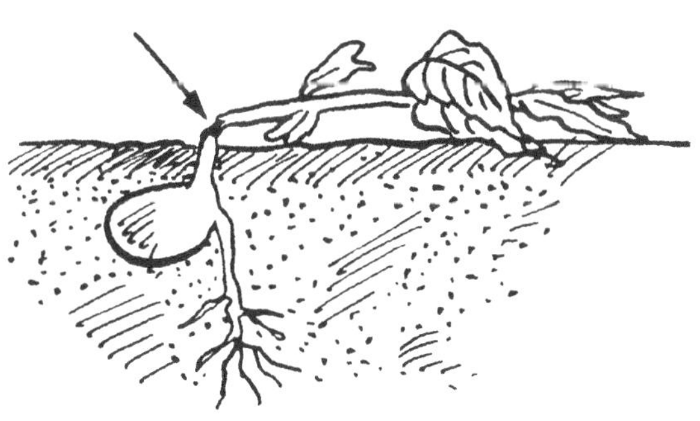

"Damping off"

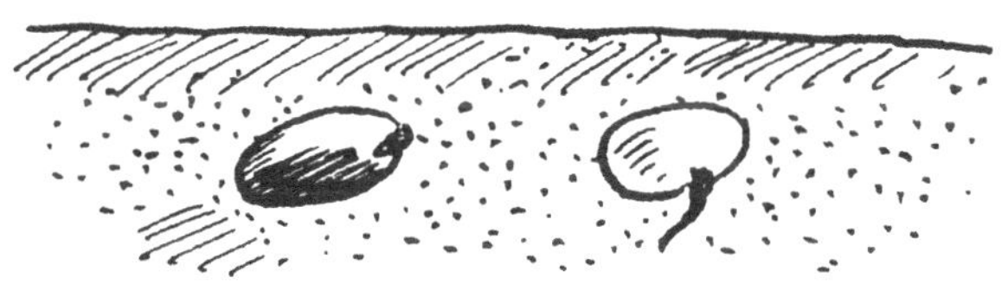

Seed rot Root rot

3. GROWING SEEDLINGS IN CONTAINERS

OPERATION 7: CARE AND TENDING

Good care can avoid most problems.

Chemicals against weeds, pests and diseases should only be used if absolutely indispensable. Remember these chemicals are strong poisons also for human beings and domestic animals!

Root (and possibly shoot) pruning

Some time after pricking out (one to four months depending on climate and species), roots start to grow out of the bottom of the pots or the drainage holes. These roots have to be cut from time to time, roughly every 2 months.

When the roots have only just started to penetrate the soil under the pot, they can be cut by simply lifting the pot. This method is known as "shocking". If the roots are too strong already, they have to be cut with pruning shears or knives. A very economical method is to pull thin piano wire underneath the complete potbed without having to lift the pots.

By using a conveniently shaped shovel or machete the roots can be cut without lifting the pots.

Root pruning should be done repeatedly rather than waiting until the roots have grown too much. Take pots from all beds from time to time to check. Initially, root pruning reduces the ability to absorb water. It should therefore be done in the evening or on a cloudy day.

In contrast to root pruning, shoot pruning is not a standard nursery operation. It may become necessary, however, if plants are overgrown due to bad planning or a delay in the planting season.

If shoots are more than twice as high as the pots are long, they can be trimmed. Even if few or no leaves remain, the seedling will survive better and establish itself faster than an unpruned one. However, shoot pruning is only possible for broadleaved species which regenerate readily, e.g. *Eucalyptus*, *Leucaena*. It should not be done for conifers or species that become bushy rather than developing one leading shoot soon after pruning.

ROOT PRUNING OF POTTED SEEDLINGS

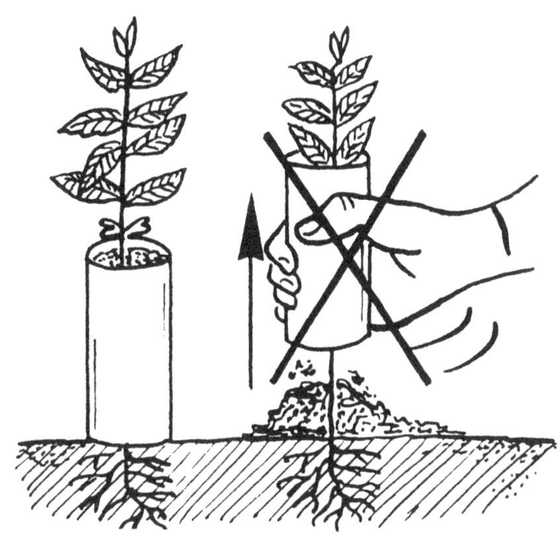

Correct: timing permits "shocking"

False: roots are too long

Root pruning with pruning shears

Root pruning with trowel

Root pruning with wire (e.g. piano wire)

3. GROWING SEEDLINGS IN CONTAINERS

OPERATION 7: CARE AND TENDING

MOST COMMON ERRORS

Watering

1. Watering is done with cans or hoses not fitted with fine nozzles. The soil is washed out of the seedbed or the pots and the seeds or roots are exposed. This causes serious damage.

2. Watering is too superficial and the lower part of the pots is not reached. Roots cannot develop there, seedlings grow more slowly and are of a lower quality.

3. Watering is too frequent and too much. Leaves or needles are yellowish because of lack of air (oxygen) for the roots. The shoot grows very fast but is soft. The shoot is not resistant and too big compared to the root. Survival in the field will be low.

Weeding

4. Weeding starts only when weeds are already tall. It requires much more time and the roots of the seedlings are damaged when pulling out the weeds.

Shading

5. Shading is too dense and/or applied for too long. Seedlings grow slowly and their shoots are long, thin and have few leaves.

6. Shading mats placed too low. After germination in seedbeds or over potbeds the shoots are deformed when touching the shading mats.

Root pruning

7. No or late root pruning means that all hair-roots, the fine roots that absorb water and nutrients, have grown into the soil underneath the pot. They are cut when lifting the seedlings for out-planting. The remaining bigger roots cannot feed the seedling and survival rate will be low.

CARE AND TENDING - MOST COMMON ERRORS

Watering

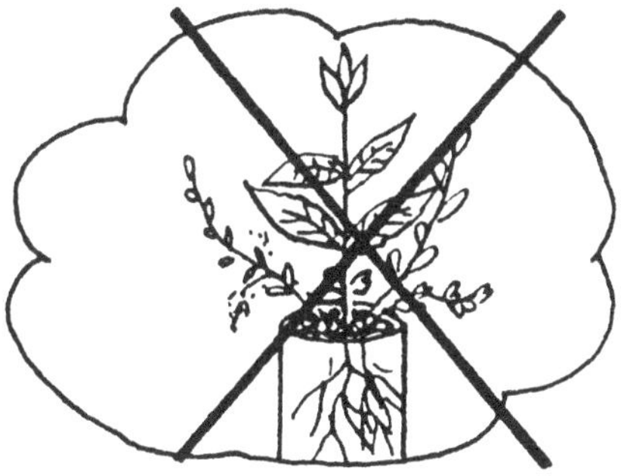

Weeding

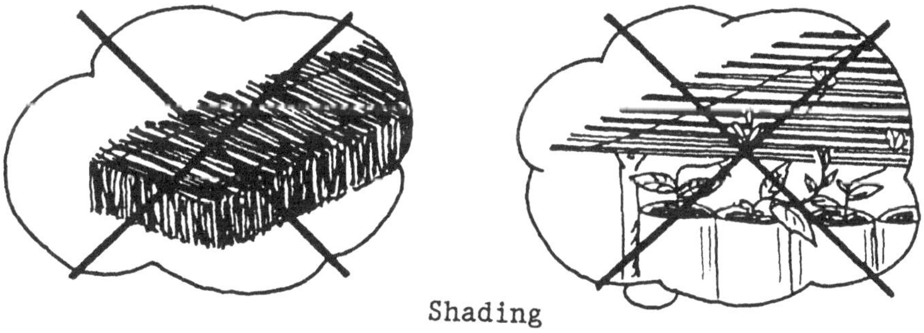

Shading

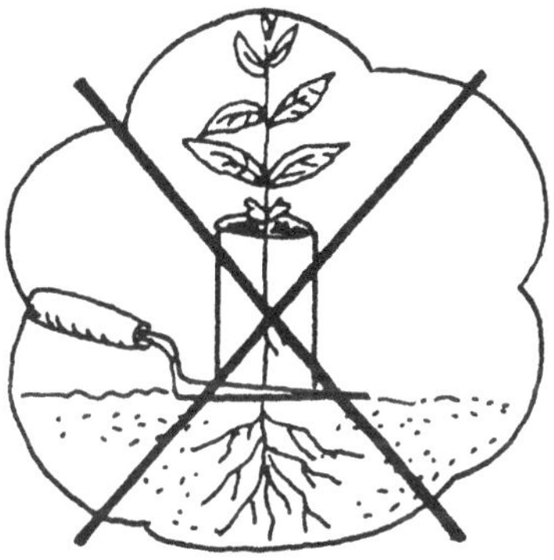

Root pruning

3. GROWING SEEDLINGS IN CONTAINERS

OPERATION 8: PREPARATION FOR PLANTING OUT

Before the seedlings can be delivered for planting in the field, some steps are still required:

1. Hardening-off
2. Grading
3. Packing.

Hardening-off

Seedlings are kept under very favourable conditions in a nursery. In the field, this will no longer be the case and seedlings have to be adapted to harsher conditions, particularly the absence of regular water supply. This adaptation is achieved by exposure to full sunlight and a gradual reduction of watering frequency starting one or two months before out-planting. The change has to take place step by step, not abruptly.

This process will make the stems hard and woody, the crown relatively short but vigorous and the root system compact and well developed.

Plants have to be watched carefully during this process. If signs of wilting appear (other than a temporary one in the heat of the afternoon), the plants have to be watered and the reduction has to be done more slowly.

Grading

Good seedlings have:

- a shoot between one and two times the length of the root (i.e. the pot);
- a sturdy, woody stem with a strong root collar;
- a symmetrical, dense crown;
- a root system with many, thin roots in addition to the tap root.

Key to drawing opposite:

(a) Deformed
(b) Too small
(c) Too few needles/leaves
(d) Two-pronged
(e) Dry top
(f) Yellowish or pale needles/leaves
(g) Too small needles/leaves
(h) Too big

(i) Hair roots insufficient
(j) Too poor roots
(k) Diseases of roots
(l) Spiral growth

GOOD QUALITY PLANT

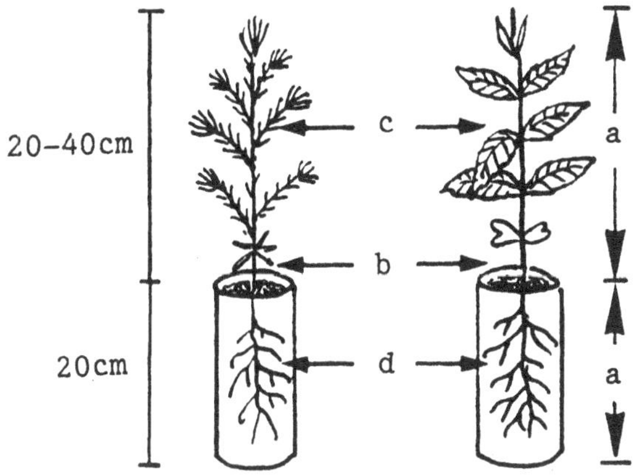

20-40cm

20cm

(a) Root: shoot from 1:1 to 1:2

(b) Woody stem with strong root collar

(c) Symmetrical dense crown

(d) Dense root system

PLANTS TO BE REJECTED

Shoots

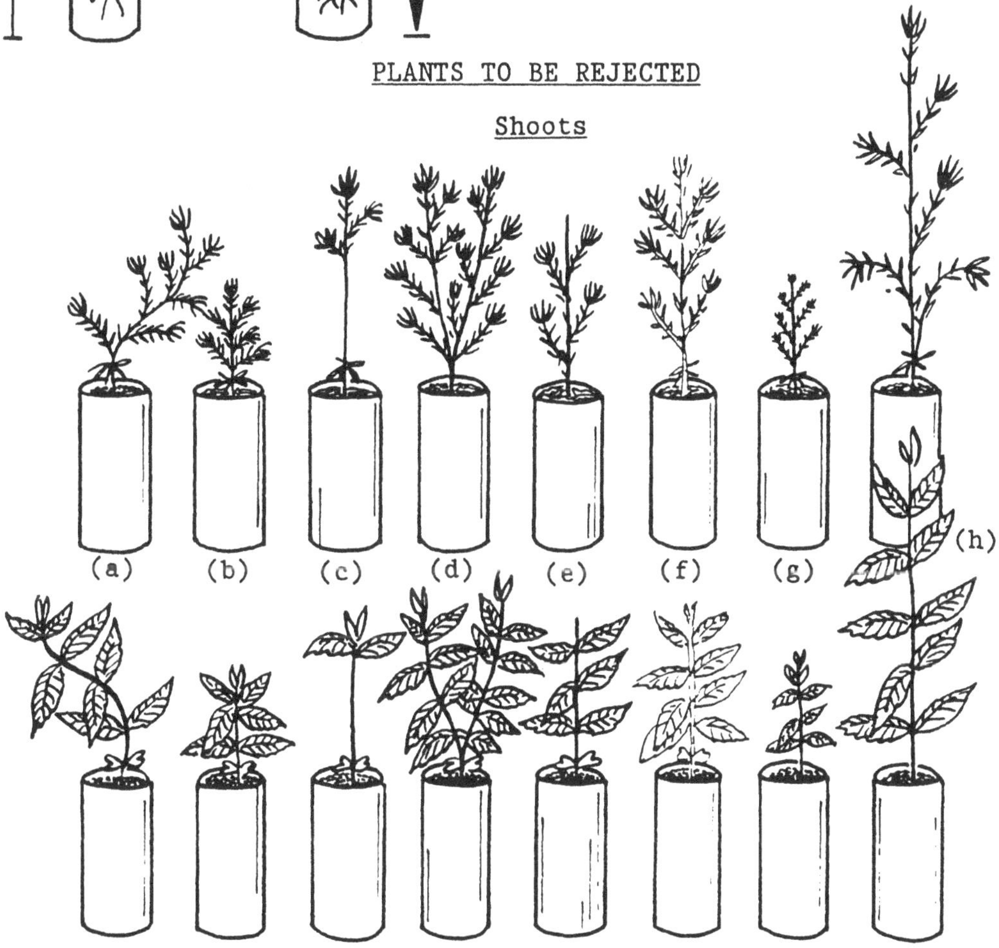

(a) (b) (c) (d) (e) (f) (g) (h)

Roots

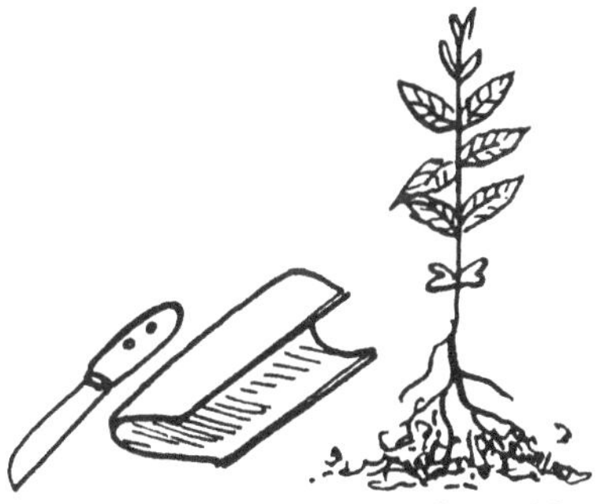

(i) (j) (k) (l)

Remove pot and soil

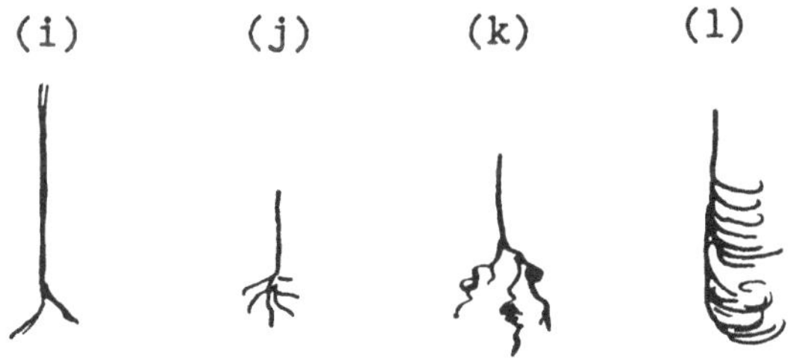

3. GROWING SEEDLINGS IN CONTAINERS

OPERATION 8: PREPARATION FOR PLANTING OUT

Only good quality planting stock should be used because planting and, even more, replacement after failure ("beating-up") is much more expensive than seedling production. Even with careful nursery work, some plants are not good enough for planting. They may be too small, have deformations or be unhealthy.

Examples are shown in the picture on the previous page. Such plants have to be eliminated and destroyed (chopping for compost after removal of the pot).

Once plants unsuitable for planting have been eliminated, further grading may be useful to separate two or three quality classes. The best plants should be used on the most difficult, inaccessible or most important sites. Acceptable, but lower quality, plants are used on more favourable sites or where replacement planting is not as costly.

The seedlings should be thoroughly watered the day before leaving the nursery unless planting takes place during the dry season or with very unreliable rainfall.

Packing and transport

Potted seedlings should be carried holding only the bags. They should never be pulled by the shoots.

Whenever possible, use boxes for transport.

PREPARATION FOR PLANTING OUT

Grading

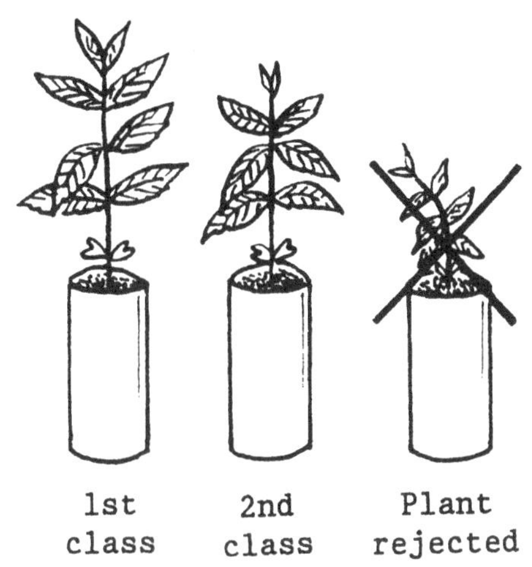

1st class | 2nd class | Plant rejected

Watering

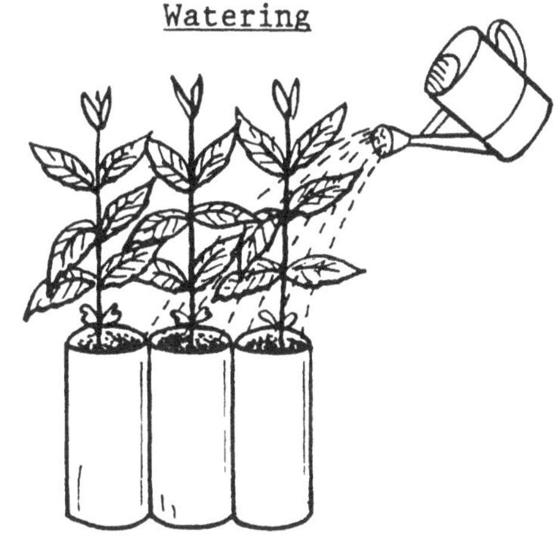

The night before planting in the field, water the seedlings thoroughly

PACKING

Hold the bags and not the seedlings

Never pull by the shoots

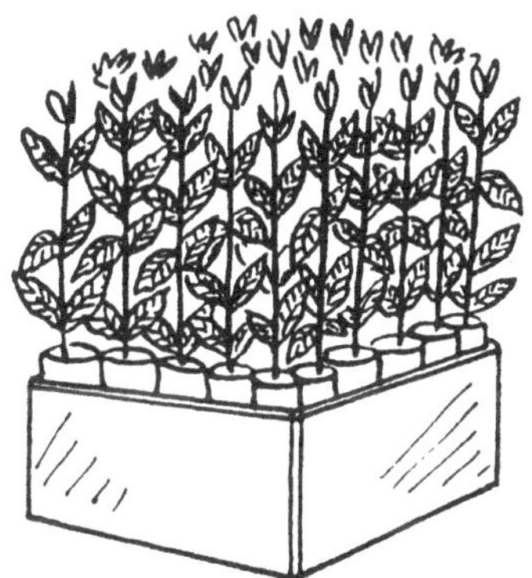

Use trays and boxes for transport

Don't tie the pots with rope

3. GROWING SEEDLINGS IN CONTAINERS

OPERATION 8: PREPARATION FOR PLANTING OUT

If the seedlings are loaded on to carts, pick-ups or trucks, load densely and upright. Never lay the seedlings down. Make sure they do not fall during transport. Metal platforms of vehicles often get very hot: this burns the root tips at the bottom of the pots. Pour water over the platform and/or spread soil, straw or twigs.

MOST COMMON ERRORS

Hardening-off

1. No, or insufficient, hardening-off means that plants do not resist harsh conditions on the planting site or delays and errors during transport and planting. Survival rates will therefore be low.

Grading

2. One of the most common and most serious errors is to only work for a high number of seedlings because targets have been fixed in quantities. Quality is just as important and normally a certain percentage of the seedlings has to be rejected and destroyed.

Packing and transport

3. Pulling seedlings by the shoots often breaks them, leading to deformation of the future trees.

4. Pots are often not placed upright and/or not secured against toppling over. The earth-balls around the roots break up and therefore the advantage of expensive containerised seedlings is lost at the last moment.

TRANSPORT

Pour water over the platform ...

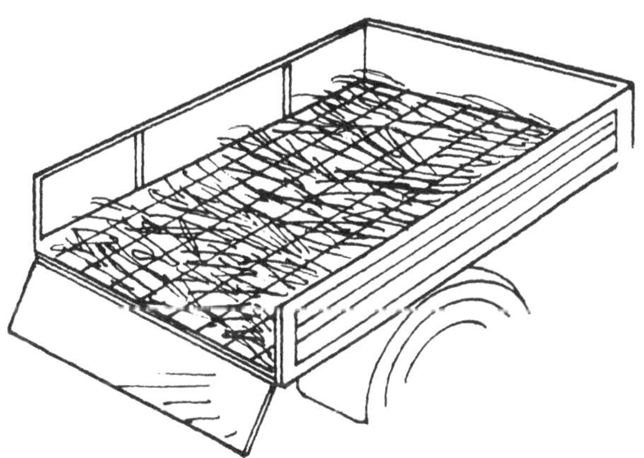

... or spread straw or twigs

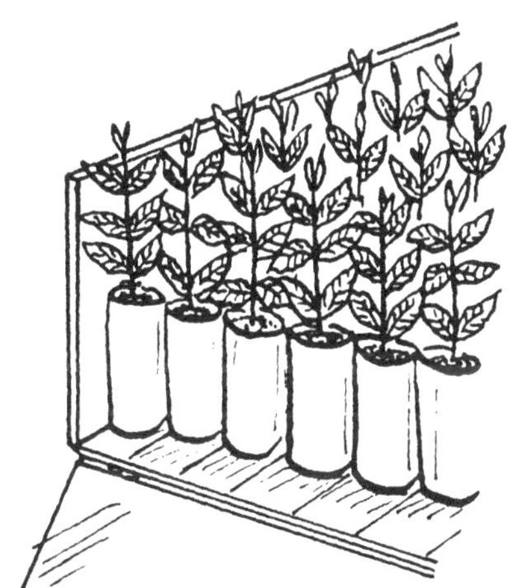

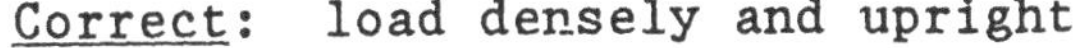

Correct: load densely and upright

False: plant lying on platform

4. GROWING BARE-ROOTED SEEDLINGS

4. GROWING BARE-ROOTED SEEDLINGS

Seedlings with bare roots, i.e. without a protective earth-ball around the roots, are cheaper to produce and to plant. Where the climate and the sites are favourable, and planting and transport reliable, this method is used. Some species like *Cassia*, *Gmelina* and *Azadirachta* are often produced successfully as bare-rooted plants even in dry climates. Important differences in the production compared to containerised seedlings are:

1. The soil: the seedlings grow directly in the nursery soil. Therefore, it has to be suitable (see Chapter 1) and has to be cultivated like a vegetable garden to a depth of 30-40 cm. Do not mix upper soil layer with lower while cultivating but keep topsoil on top of the bed.

2. Density and area requirement: Plants in the transplant beds are in rows 20-25 cm apart and the distance between plants is 5-10 cm. There are 50-100 plants per m^2 bed area, e.g. fewer plants than when using common container sizes. A larger nursery area is therefore required.

3. Transplanting: Plants to be transplanted into beds should be 6-8 cm tall, i.e. bigger than those pricked out into pots, and have a good root system (tap root 4-6 cm long).
   Transplanting is carried out by preparing holes with a dibble (or a ditch with a wedge spade), inserting the plants without twisting the roots and filling the hole with soil from the seedbed which is gently compacted around the plants.

4. Root pruning: This is necessary as for potted plants but carried out with the plants in the ground by inserting a sharp spade or machete from the sides of the row to prune the tap root and between plants in a row for the lateral roots. Root pruning may be needed every six weeks after transplanting.

DENSITY AND AREA REQUIREMENTS

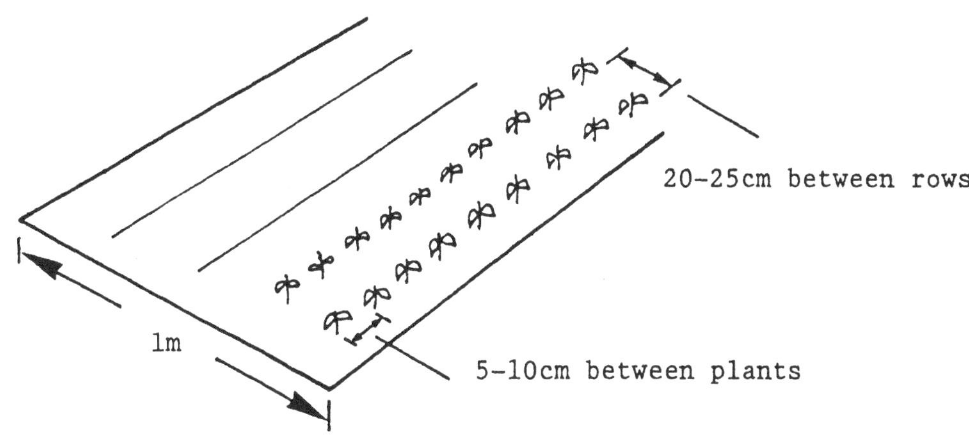

TRANSPLANTING

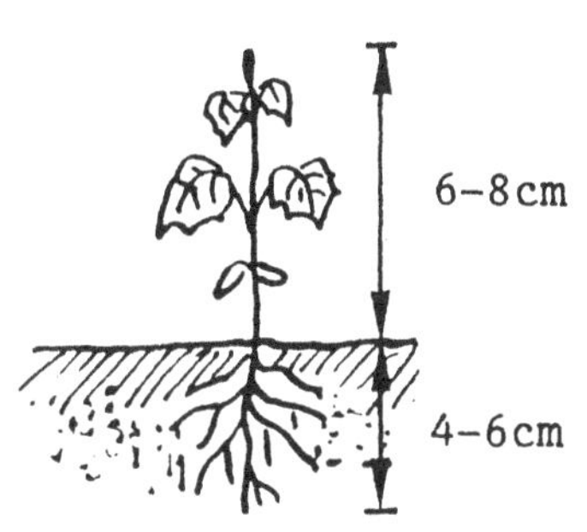

Right size of plants

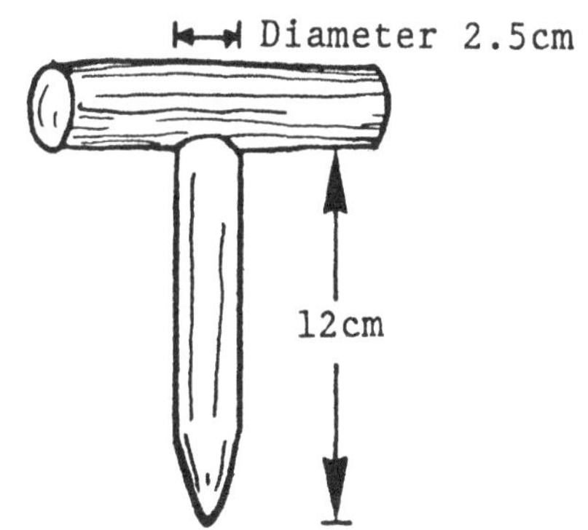

Dibble for transplanting bare-rooted seedlings

ROOT PRUNING IN A TRANSPLANT BED

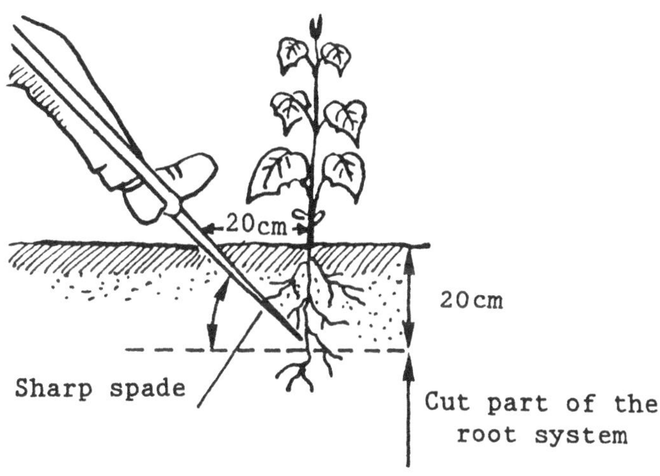

Pruning of the taproot

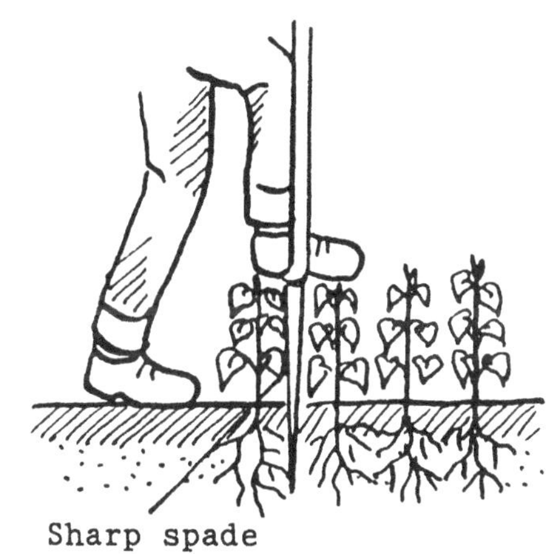

Pruning of lateral roots within the row

4. GROWING BARE-ROOTED SEEDLINGS

---

5. Lifting: A spade or a flat-pronged fork is inserted vertically at about 10 cm from the plant. It is pushed deep enough to permit lifting the plant with the entire root system.

6. Packing: After grading (and possibly root pruning), plants are packed. The roots have to be well protected against drying. Sacking, banana leaves, plastic bags with ventilation holes or cans may be used. To preserve moisture, the roots should be covered with wet grass, leaves, sawdust or mosses, where available.

   Other operations are similar to those for containerised stock.

LIFTING OF PLANTS

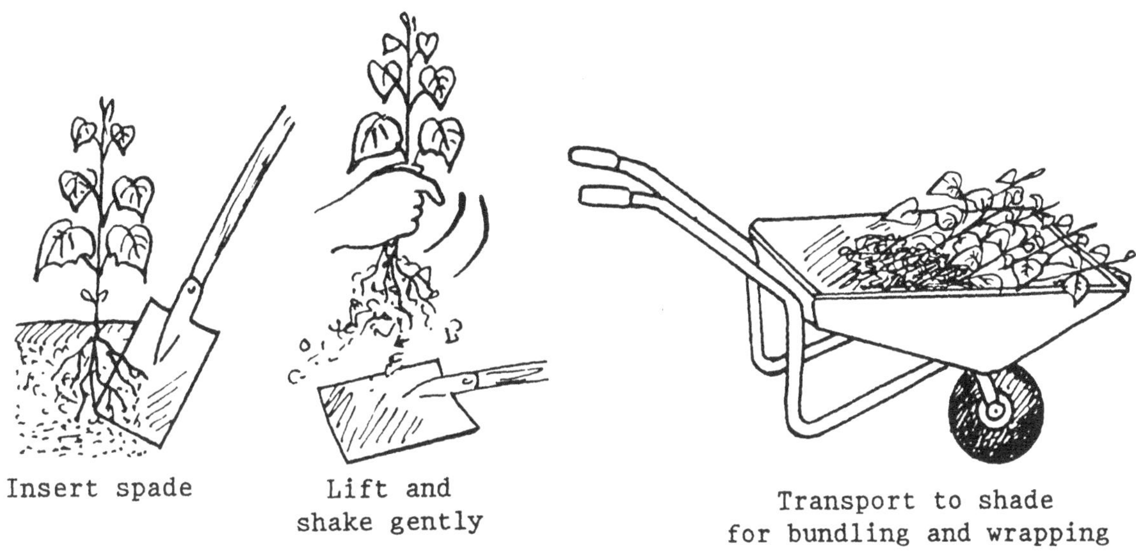

Insert spade

Lift and shake gently

Transport to shade for bundling and wrapping

WRAPPING AND PACKING OF BARE-ROOTED SEEDLINGS

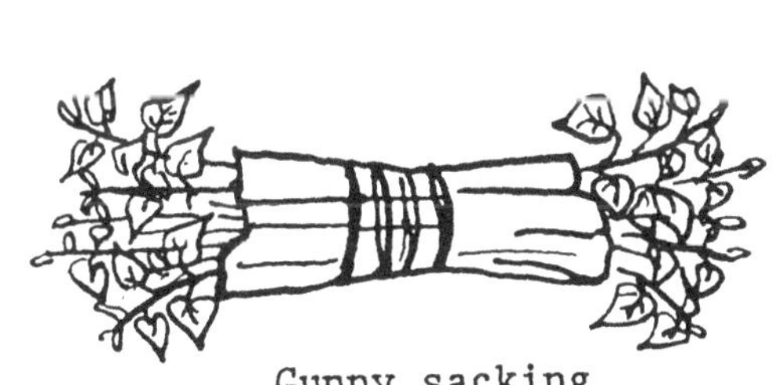

Gunny sacking

Banana leaves

Polyethylene bag with ventilation holes

Can

5. STUMPS

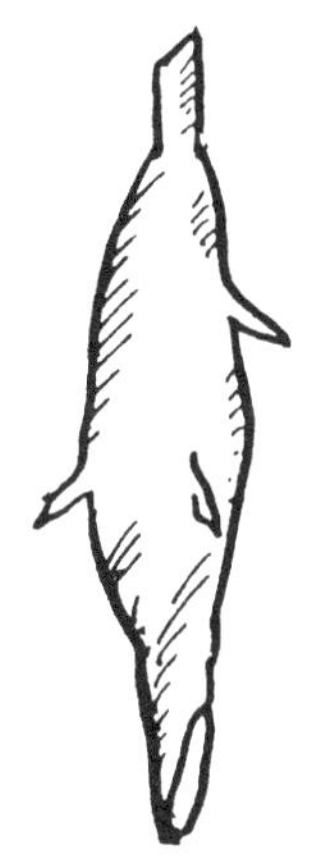

5. STUMPS

"Stumps" are seedlings of which most of the shoot and side roots have been removed, leaving only a 20-25 cm long piece.

Stumps can be planted successfully only for a limited number of broadleaved species. Most commonly stumps are used for Teak (Tectonea grandis) but also for Gmelina (G. arborea), Cassia (C. siamea), Neem (Azadirachta indica), Pterocarpus spp. and Acacia cyanophylla.

For these species, stumps have a number of advantages:

- they are light and therefore easy and cheap to transport;
- simpler and cheaper planting techniques can be used;
- they can be cut and delivered as required;
- they can be stored for about a week without reduction in survival rates;
- they can already be planted just before or very early in the rainy season;
- for some species, they have higher survival rates than bare-rooted plants because the water-consuming shoot has been removed.

Some of these advantages are particularly interesting for nurseries distributing seedlings to rural people.

The steps to produce stumps are:

1. Grow bare-rooted seedlings to a shoot-size of 60-90 cm.
2. Lift seedlings and cut shoot 2-5 cm above root collar.
3. Trim tap root and side roots.
4. Mud-puddle by dipping, bundle and wrap (in banana leaves, for example).

The most suitable dimensions are:

	Teak	Gmelina
Root collar diameter	1-2.5 cm	1-2.5 cm
Shoot length	3-4 cm	5 cm
Root length	15-20 cm	15-25 cm
Lateral roots	removed completely	shortened to 2-5 cm

STUMPS

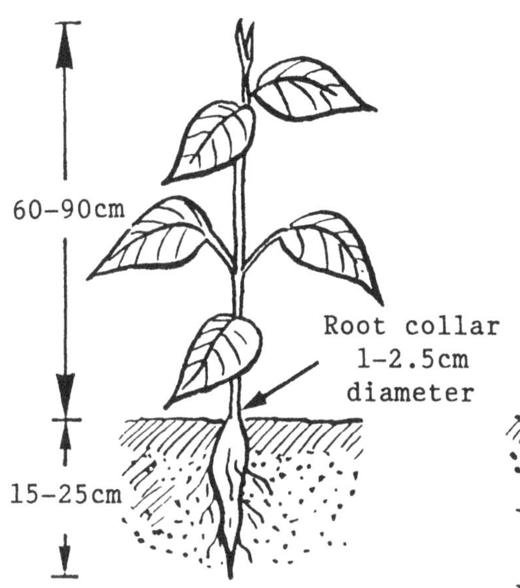

Grow plant in stock bed

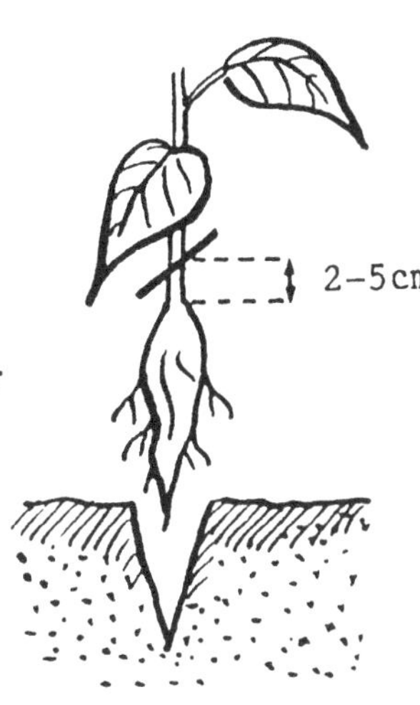

Lift seedling and cut shoot

Trim tap root and side roots

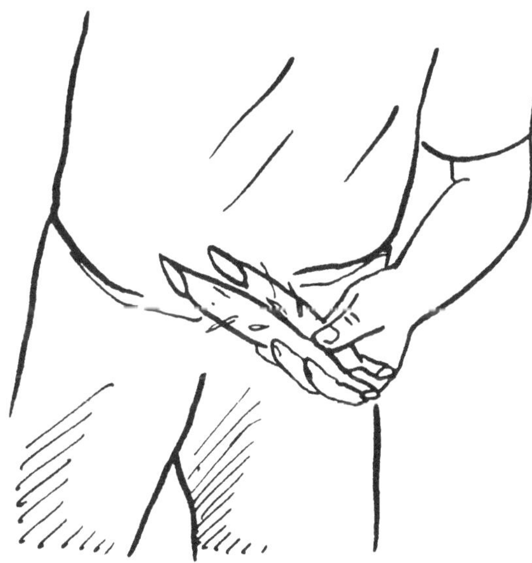

Teak "stumps" suitable for planting

Mud-puddling and bundling

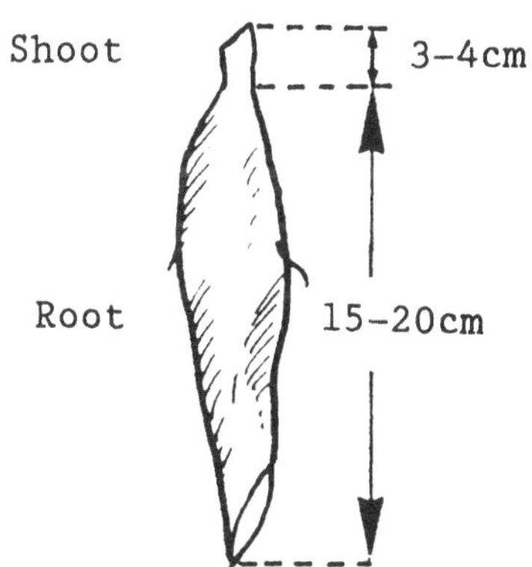

Right size teak

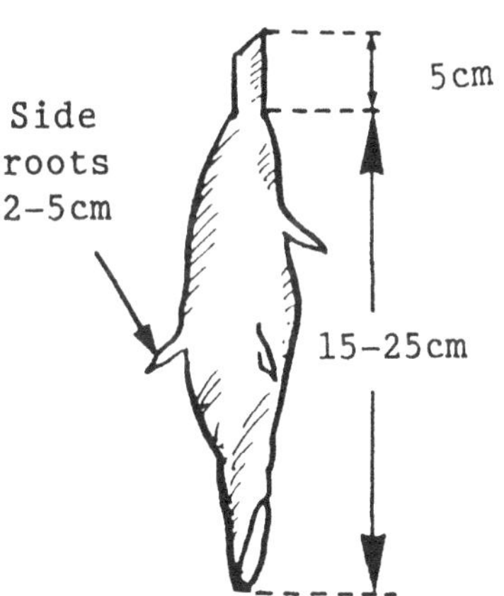

Right size gmelina

6. GROWING GRAFTED FRUIT TREE SEEDLINGS

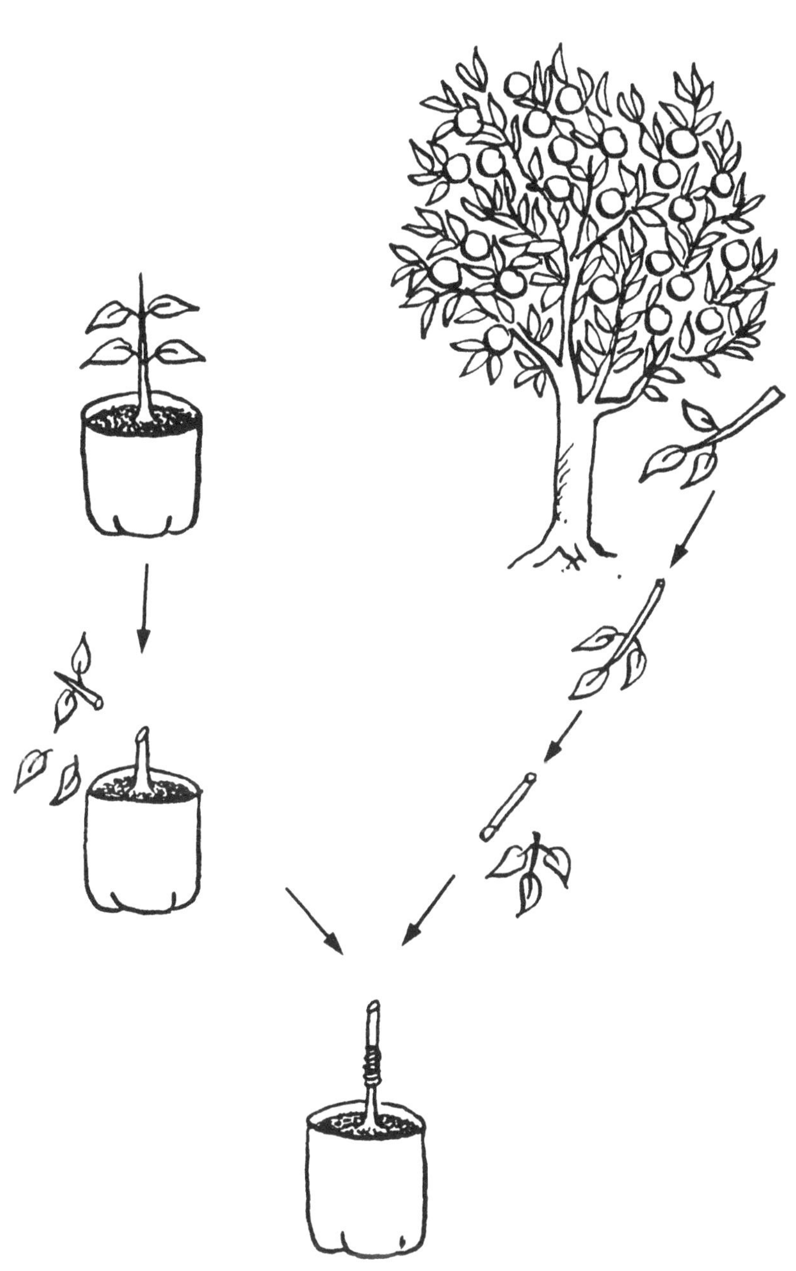

6. GROWING GRAFTED FRUIT TREE SEEDLINGS

Fruit and nut tree seedlings are often in high demand by the local population because they produce foods that add vitamins, proteins or fats to their diet which are otherwise missing. Fruits and nuts may also be attractive cash crops.

The growing of fruit and nut trees requires the same facilities and steps as that of forest trees. It can therefore easily be included in the production programme of a forest nursery.

The only additional activity which should be carried out for fruit and nut trees is grafting. This technique is explained in detail in the following.

The best way of learning to graft is to assist an experienced horticulturalist for a day. Experienced individuals capable of providing on-the-job training can often be found in agricultural research stations or specialised fruit-tree nurseries.

What is grafting?

Through grafting, branches from high-yielding fruit-bearing trees can be united with a seedling from the same or a closely-related species.

Why graft?

- Because the grafted twig is already mature to flower and bear fruit.

- Grafted seedlings of fruit and nut trees start to bear fruit much earlier: 2-5 years after planting instead of 10-15 years.

- Grafted seedlings can bear more tasty fruits and nuts than local varieties but at the same time retain the resistance against pests and diseases and the adaptation of their rooting system.

GRAFTING

GRAFTED

Seedling with graft → 3rd year → 10th year

NOT GRAFTED

Seedling → 3rd year → 10th year

6. GROWING GRAFTED FRUIT TREE SEEDLINGS

How to graft

1. Produce seedlings of the desired species in larger containers (20-25 cm flat) with a bottom. Alternatively big enough transplants can be used. Use seed from local varieties well adapted to climate and soil and also resistant to disease.
   The trees from which seed is collected have to be healthy and vigorous; the quality of their fruits does not matter.

2. The seedlings are ready for grafting when their diameter at 40-60 cm height has approximately reached that of a pencil.

3. The best time for grafting is the second half of the dry season.

4. The twigs to be grafted on to the seedling (called "scions") are collected from trees with a lot of good-quality fruit, such as genetically improved varieties. Fully mature, i.e. woody, one-year-old twigs, with approximately the diameter of a pencil are collected shortly before the grafting is to be carried out to avoid their drying out.

5. For scion and seedling to unite, it is essential that a firm contact between the growing zones of the two is established. This zone is found just under the bark.

HOW TO GRAFT

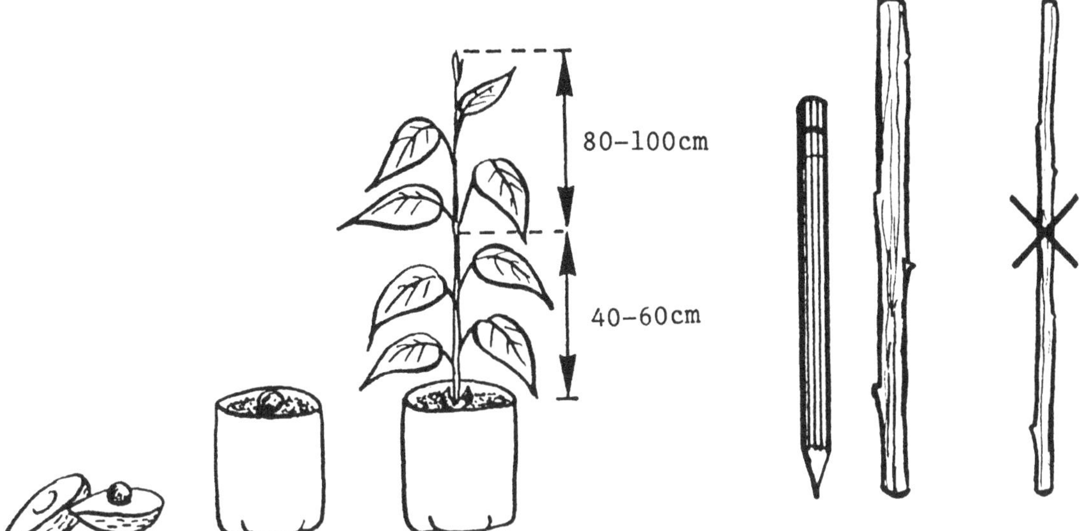

Seedlings are ready for grafting

Correct: diameter of a pencil

Oblique cut on the scion

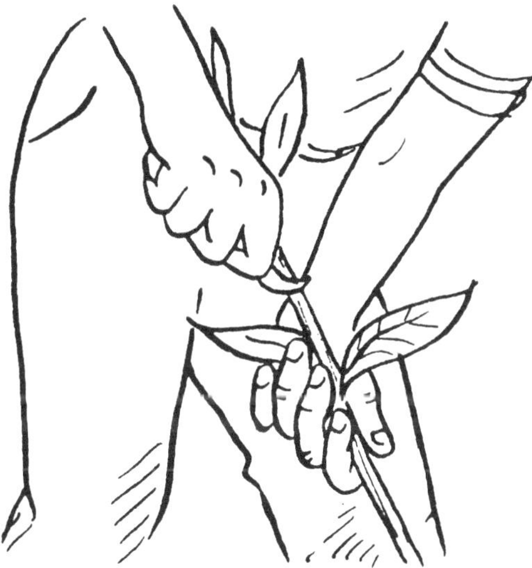

Oblique cut on the seedling

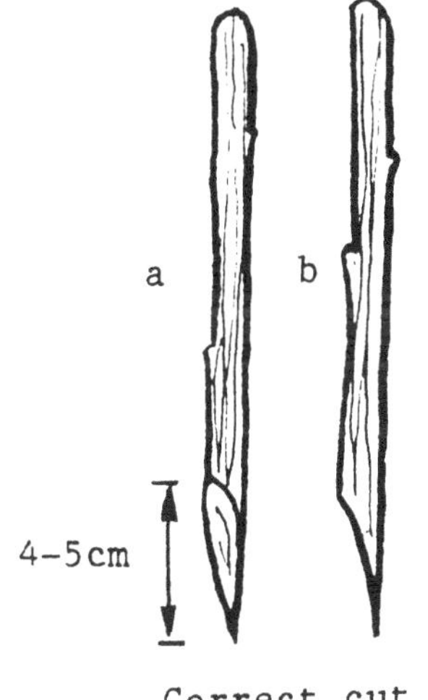

Correct cut

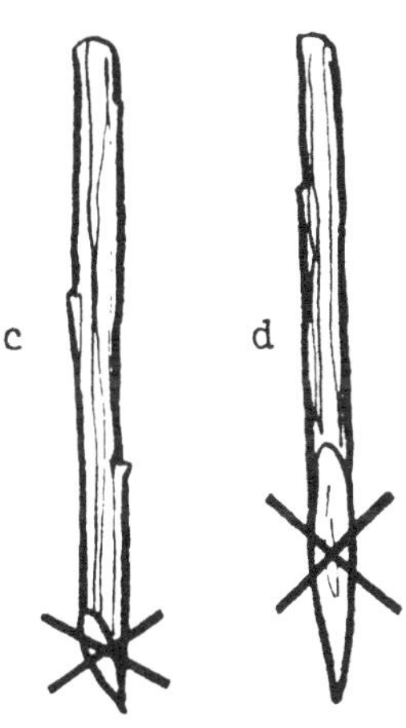

False

6. GROWING GRAFTED FRUIT TREE SEEDLINGS

---

6. Two methods of grafting are common: one where oblique cuts are made on both scion and seedling and the two surfaces laid one on to the other. the second is a slot in the seedling and two cuts on the scion.

   If both seedling and scion have the same diameter, either of the methods can be used. If the seedling is stronger than the scion, method 2 (the slot) is preferable.

   Other, more complicated methods of shaping scion and seedling end are also practised by horticulturalists.

7. Correct cutting is very important. As shown on the preceding page, the blade of the knife is placed almost parallel to the twig. A gently pulling cut, in one go, should produce a completely level surface of about 4-5 cm length. If the cut is shorter, scion and seedling will not join well. Long cuts from repeated cutting have an uneven surface and do not join well either.

8. The cuts should be on the side of the twig opposite to the next bud. Four to five buds should remain on the scion.

9. The surfaces of the cuts on scion and seedling are then put on to one another. Make sure the growing areas of both can establish firm contact on at least one side.

10. Seedling and scion are tied together with bast, woollen thread, strips of leftover polyethylene or the like. Take care not to change the position of the sufaces.

11. Until the join is completed, grafted seedlings should be protected from intensive sunshine and watered regularly.

HOW TO GRAFT

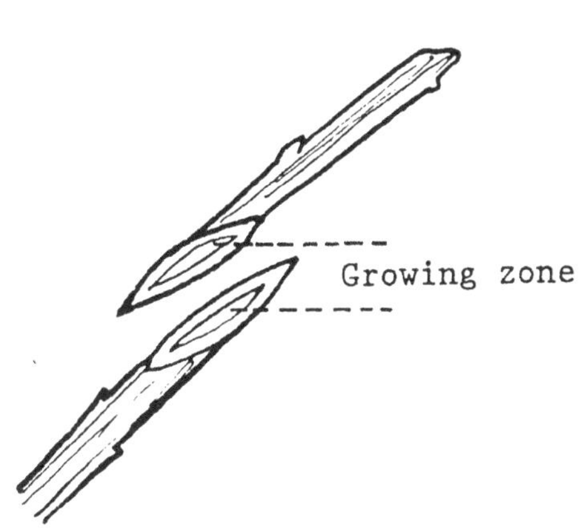

Method 1: sideways

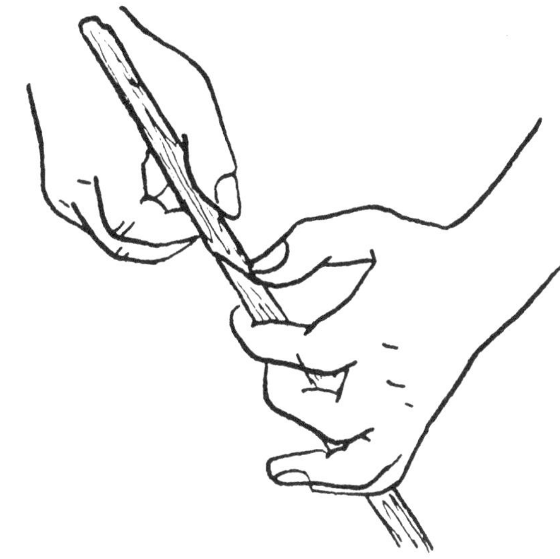

The two surfaces are laid on to each other

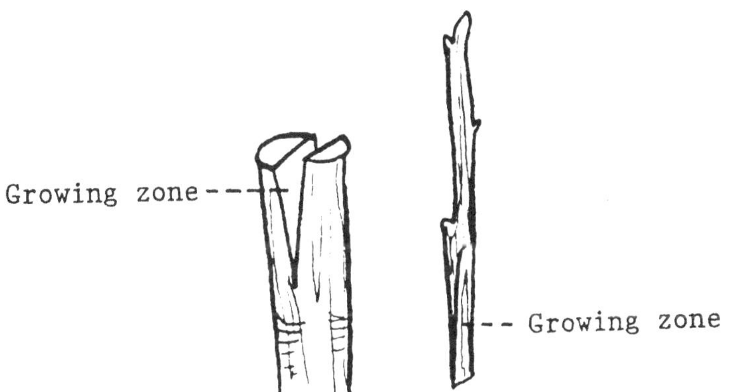

Method 2: slot

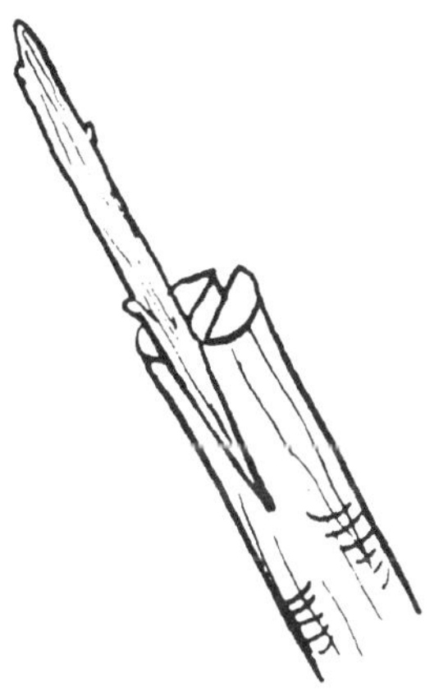

Correct: firm contact of two growing zones on one side

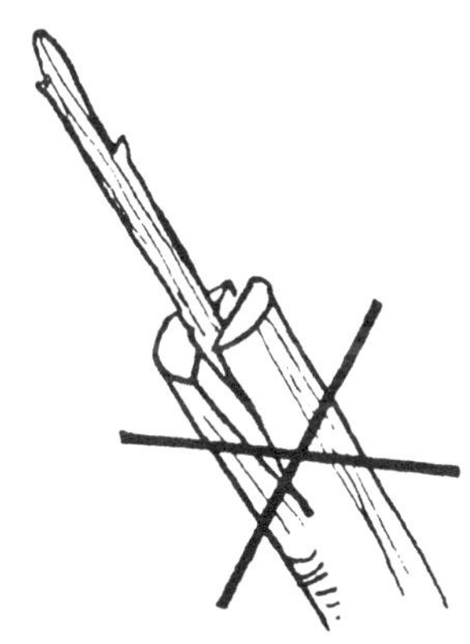

False

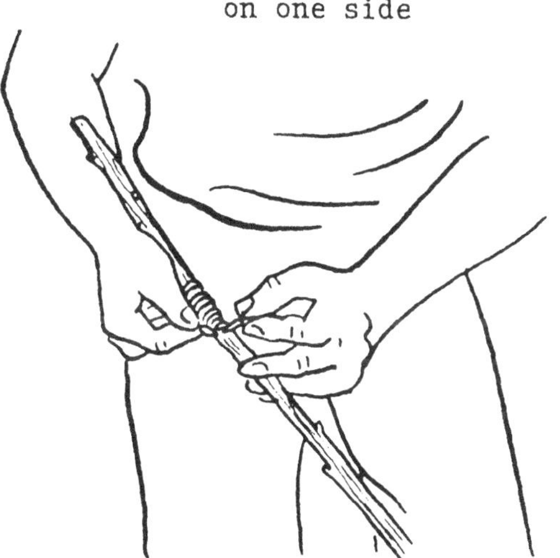

Scion and seedling are tied together

7. NURSERY PLANNING, WORK ORGANISATION AND RECORDING

7. NURSERY PLANNING, WORK ORGANISATION AND RECORDING

7.1 PRODUCTION CALENDAR

For successful tree planting, seedlings of the right species, size and of good quality are essential. Otherwise, survival rates are low and the cost of planting, which is much higher than that of seedling production in the nursery, is wasted.

Nursery work is therefore one of the forestry activities requiring particularly careful planning. It has a fixed deadline - the best time for planting. Even during the production of plants in the nursery, good timing is very important as mistakes are difficult or impossible to rectify (e.g. pricking-out or pruning roots too late).

To make sure all operations are carried out on time, the following are needed:

1. A production calendar showing when the nursery operations have to be carried out for the species to be produced.

2. A schedule for the supply of materials, tools and equipment.

3. The right kind and numbers of workers organised in a suitable way.

4. Nursery records on total production, timing labour and other inputs used.

Since the planting season is the deadline, the planning for nursery activities has to be calculated backwards from this date:

date of planting
minus number of days for growing from pricking-out to planting size
minus number of days from germination to pricking-out
minus number of days from sowing to germination
equals sowing date.

Example for calculation of sowing date

Assumed planting dates 1 November - 12 December
best date 15 November

Species A: (required in smaller numbers; planted in a short period of time)

No. of days required:
- from pricking out to planting size 150
- from germination to pricking out 20
- from sowing to germination 15

Total 185 days

15 November - 185 days = 14 May

Species B: (required in large numbers, say 50,000; planting spread over whole planting season)

No. of days required:
- from pricking out to planting size 260
- from germination to pricking out 30
- from sowing to germination 10

Total 300 days

First batch (e.g. 10,000)
Planting date 1 November - 300 days = 14 January

Second batch (e.g. 25,000)
Planting date 15 November - 300 days = 29 January

Third batch (e.g. 15,000)
Planting date 7 December - 300 days = 20 February

EXERCISE

Work out the planting dates for

Species C (larger numbers; planted in two batches)

Assumed best planting date 1 December

No. of days required
- from pricking out to planting size 214
- from germination to pricking out 30
- from sowing to germination 45

Total 289 days

First batch to be planted 25 November, to be sown? ____________________

Second batch to be planted 7 December, to be sown? ____________________

7. NURSERY PLANNING, WORK ORGANISATION AND RECORDING

7.1 PRODUCTION CALENDAR, 7.2 SCHEDULE FOR SUPPLIES

A calendar showing the duration of these stages for all important species is very helpful for good timing. An example for a highland nursery is shown opposite. In warm lowland areas, production time will be considerably shorter. In some regions, there are two rainy seasons during which tree planting is possible. In this case, there are also two nursery seasons per year. Note also that different species need different times, so they should not be sown all at the same time. If the right time for a species has been missed by more than a month, it should not be sown at all because seedlings would be too small at planting time. However, if such a situation cannot be avoided by good planning, it may be possible to select an alternative species also suitable for the planting purpose and site but germinating and developing faster. This is preferable to delivering seedlings much too small for planting.

Records should be kept about the development of species in each nursery (dates of sowing, germination, pricking out, size at planting time) so that a calendar for the local conditions can be developed.

The production calendar opposite shows when operations are due for the species to be planted and when materials, tools, etc. should be procured. The quantities required can be estimated from the production programme specifying the number and quality (and size) of the seedling to be produced. An example of the calculation of pot and potting soil requirements is given in Appendix 3.

To make sure supplies arrive on time, allowance has to be made for the request to reach the procurement officer, delays in placement of orders and delivery.

NURSERY CALENDAR - COMMUNE OF KIVU, HIGHLANDS OF RWANDA, EAST AFRICA (1900-2000 m)

(planting season end October/November)

SPECIES	Jan	Feb	March	Apr	May	June	July	Aug	Sept	Oct	Nov	Dec	Jan
Acacia melanoxylon					//////	::::::==	=====	=====	=====	=====	!!!!!		
Cedrela serrulata				////	//:::	::::==	=====	=====	=====	=====	!!!!!		
Callitris robusta	////	//:::	:::===	=====	=====	=====	=====	=====	=====	=====	!!!!!		///
Cupressus benthamii	/	/////	::::::==	=====	=====	=====	=====	=====	=====	=====	!!!!!		
Eucalyptus maideni				/	///::	::===	=====	=====	=====	=====	!!!!!		
Eucalyptus saligna				////:	:::===	=====	=====	=====	=====	=====	!!!!!		
Grevillea robusta				/////:	:::===	=====	=====	=====	=====	=====	!!!!!		
Markhamia platycalyx				/////:	:::===	=====	=====	=====	=====	=====	!!!!!		
Maesopsis eminii			/////))	)))::	:::===	=====	=====	=====	=====	=====	!!!!!		
Hagenia abyssinica			/////:	:::==	=====	=====	=====	=====	=====	=====	!!!!!		
Faurea saligna		///	//::::	=====	=====	=====	=====	=====	=====	=====	!!!!!		
Podocarpus usumbarensis	))))::	::::::	=====	=====	=====	=====	=====	=====	=====	=====	!!!!))	)/////	)))))
Polyscias fulva				//////	::::==	=====	=====	=====	=====	=====	!!!!!		
Persea americana		//	///===	=====	=====	=====	=====	=====	XXXXX==	=====	!!!!		

Key:

////	Sowing	====	Care and tending
))))	Germination	!!!!	Planting
::::	Pricking out	XXXX	Grafting

An adequate number of reliable workers is crucial for successful nursery production. Some factors have to be considered before the number of workers can be estimated:

- are permanent or temporary workers to be used?
- should they be male and/or female?
- is the work paid on a time- or on a piece-rate basis?
- should all workers do all kinds of work or should they be specialised?
- how can favourable working conditions be created that improve workers' motivation and productivity?

Permanent or temporary workers

The choice depends on the distribution of work and the qualification required.

The volume of work in a nursery is not constant over time. Some activities requiring a lot of labour have to be carried out in a short period of time, leading to labour peaks; other work is spread more evenly. In nurseries with only one season per year, there are "slack periods" when very little work needs to be done. In the example for a nursery calendar on p. 93 the busiest periods are preparation and filling of containers, and pricking-out in the months of late March, April and early May, while in July, August and September, only regular maintenance is needed, requiring less labour. After seedling delivery in November, only repair work is done until the next season in February.

Some nursery work requires skill and experience, e.g. sowing, pricking-out, grading, while some can be performed after a short initiation, e.g. preparation of potting soil and pot filling, manufacture of shading mats, etc.

Ideally, a permanent workforce should be available to carry out more demanding and routine tasks (weeding, watering, root pruning) while temporary labour is employed at peak periods to take on simpler tasks. The most important example of this is additional labour for the preparation of potting soil and the filling of pots.

The foreman and, where required, a watchman, always have to be permanent staff.

PERMANENT OR TEMPORARY WORKERS

Permanent workers

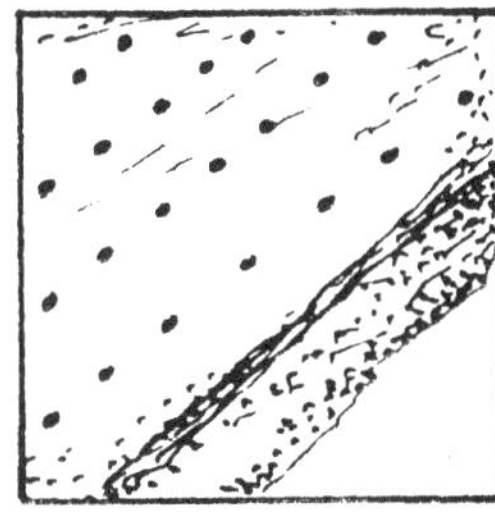

Sowing

Pricking out

Grading

Foreman

Watchman

Temporary workers

Pot filling

Manufacture of shading mats

7.3 WORKERS AND WORK ORGANISATION

Male or female workers

Experience in many countries shows that women are better nursery workers than men. Very few nursery tasks demand heavy work but many require care, patience and nimble fingers. To the extent that local culture and society permit, women should be preferred to men. Likely exceptions are the post of the watchman and workers for tasks like construction, potting soil procurement and tool maintenance.

Wages on a time or piece-rate basis

The best system is probably a mixture of both forms of remuneration. Some tasks lend themselves well to piece-rate work: preparation of pots and potting soil, the filling of pots and the manufacture of shades. Quality standards can easily be established and checked and no particular care is required for the work. Which piece-rates to use is difficult to say in general. The figures in the table on page 103 may be used for a first approximation if no local experience is available. Other operations should never be paid on a piece-rate basis because care is needed and the effects of rushed work often only show much later. Such operations include sowing, pricking-out, weeding, watering and grading. For such operations and some other tasks, e.g. tool maintenance, compost making, keeping the nursery clean, the payment of a bonus can help to improve performance. This bonus is a fixed amount of money paid in addition to the normal salary or wage if a clearly defined quality and quantity of work has been accomplished (e.g. certain number or percentage of grade seedlings, or weed-free nursery).

WAGES

WAGES ON A TIME BASIS

From  to 

Sowing
weeding
pricking out
watering
grading

WAGES ON A PIECE-RATE BASIS

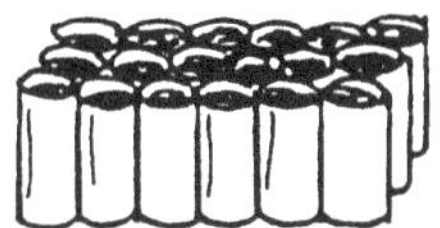 = 

Preparing pots
Filling pots
Manufacture of shades

BONUS

No bonus

Bonus

Specialisation

While permanent nursery workers have to be flexible and perform a number of different jobs in the course of the production season, the assignment of certain tasks to individuals can improve performance and make supervision easier. Such tasks are sowing, grading, compost making, tools and equipment (e.g. irrigation pump) maintenance, nursery cleaning. Such assignments also make the use of a bonus easier and more effective.

Working conditions

It is often overlooked that productivity can be raised substantially through relatively small investments into improving working conditions. In nurseries one should provide the workers with shade or shelter against rains and strong winds to avoid the negative impact these would otherwise have on the workers' performance and well-being. Safe drinking water should be available in sufficient quantity, especially during hot weather. If water supplies permit, workers should be allowed to use nursery water for washing and showers. A fireplace should be available to prepare meals and hot drinks, as well as for the night-watchman.

The provision of simple appliances like a small stool for weeding or a yoke for water carrying can greatly reduce the work strain. Tools of the right size and weight (e.g. for women, smaller and lighter than for men) and well maintained are also important. It is also a good idea to permit the foreman or permanent workers to grow vegetables in an unused part of the nursery. Their more frequent presence, including after working hours and during the week-end, will benefit the nursery. Where possible, a permanent worker should live in or very close to the nursery.

EXAMPLES OF SPECIALISATION

WORKING CONDITIONS

Safe drinking water

Fireplace

Stool for weeding

Corner for gardening

The number of workers required

The size of the workforce required depends on many factors, including the number of seedlings to be raised, the length of the production time, climate, species, whether they are containerised or bare-rooted, the type of irrigation used (watering cans, sprinklers, channels), the length of the workday and others. The table on the next page indicates an order of magnitude.

To give an idea of the labour input so as to allow for comparison with other nurseries, the number of workdays in a season is divided by the number of plants produced in thousands (e.g. 1,380 workdays to produce 114,750 plants; 114,750 rounded to the nearest thousand = 115,000; 1,380 : 115,000 = 12 workdays/1,000 plants) or the number of plants produced per day (1,000 : 12 = 83).

Nursery costs represent a considerable part of the overall cost of a plantation. The person in charge of the nursery therefore has to be cost-conscious.

Apart from the costs for pots and sometimes for transport or irrigation, labour is the most important cost item in a non-mechanised nursery.

Figures from Special Public Works Programme projects range from 10-40 workdays but could probably be improved to range from 5-25 with most between 10 and 15 workdays per 1,000 plants.

If no local experience is available, a permanent workforce of 1 foreman, 5-6 workers, 1 watchman can be reckoned with for the operation of a nursery producing about 100,000 containerised seedlings with only one planting season per year. This assumes additional temporary labour (300-400 workdays) for pot-filling, etc.

ORDERS OF MAGNITUDE FOR WORK INPUT LIKELY TO BE REQUIRED IN SPECIAL PUBLIC WORKS PROGRAMME NURSERIES

(non-mechanised, mixed species, production time 7-8 months)

Type of work	Productivity			
	Output/workday		Workdays/1,000 plants	
	Range	Most common	Range	Most common
Preparation of beds and sowing			0.7- 2.4	1.2
Preparation and filling of containers	650-185	370	1.5- 5.4	2.7
Pricking-out	830-385	770	1.2- 2.6	1.3
Watering	5000-200	385	0.2- 5.1	2.6
Weeding	3300-180	435	0.3- 5.6	2.3
Root pruning	5000-590	1250	0.2- 1.7	0.8
Lifting and loading	2000-770	625	0.5- 1.3	0.6
Miscellaneous	3300-1000	500	0.3- 1.0	0.5
Total			4.9-25.1	12.0

Note: The above are not "standard" work norms but are meant to give an idea. However, if the "most common" figures are not reached, for which there may be good reasons, project management should ask why.

To provide information about the performance, cost and productivity of the nursery, as well as to improve planning and operations in the future, a number of records have to be kept in the nursery.

These include:

- worker attendance sheets and pay rolls;
- store inventories;
- delivery reports;
- plant development records; and
- nursery inventories.

Only the last three will be considered here because forms and instructions for the others are usually provided by project management.

Delivery report

This report summarises how many seedlings of each species left the nursery, their quality, the dates of delivery, who received the plants and to which site they were sent. The recording of the recipients is particularly important if seedlings are distributed to the rural people or institutions so that later the survival of the seedlings can be checked. An example for a form to use is shown on the next page.

DELIVERY REPORT

sery: ______________ Season: ______________ Foreman: ______________
(name/location) Supervisor: ______________

	Seedlings delivered			Date	Recipient Name, address	Planting site	Remarks
	Grade 1	Grade 2	Total				
ecies 1							
ecies 2							
ecies 3							
ecies 4							
c.							

Plant development record

Successful nursery work depends to a large extent on experience with the behaviour and particular problems of each species produced under the conditions of the nursery in question. To make sure all information is collected to enable new personnel to benefit from the experience already acquired, systematic recording in writing is recommended.

After a few seasons, it will be possible to see from these records:

- how long it takes for a species to grow to the right planting size;
- how much seed is needed to produce the required number of plants;
- how long the seed of a species takes to germinate and whether it germinates all at the same time or over a longer period. This helps to plan the labour requirements over the season;
- which species need additional fertiliser, more or less watering and shading, special precautions against pests or diseases.

The form opposite can be used for this purpose.

Once timing for good seedling development is known, it is also easy to plan preparatory work (obtain soil mixture, fill pots, etc.) in time to place orders (new pots, tools, etc.) and to see when there are particularly busy weeks (pot filling, pricking-out) for which enough labour has to be provided.

PLANT DEVELOPMENT RECORD

SPECIES ____________________

Seed source: ____________________

SEED BED NUMBER ____________________
(or pot bed number
(for direct sowing)

SOWING

Date:

Date germination starts:

Remarks on germination:

Quantity seed:

Pretreatment:

Date germination completed:

PRICKING-OUT

Date:

No. of plants pricked out:

Remarks:

Potbed number:

REMARKS ABOUT PLANT DEVELOPMENT:
(disease, weeds, fertiliser, watering, etc.)

PLANTING SEASON

Date ready for planting:

Number of plants delivered: Total: Grade 1: 2: 3:

Height smallest plant delivered:

Height tallest plant delivered:

Average height:

Number of plants rejected:

Number of plants left:

Location of planting site:

Nursery inventory

For the planning and preparation of planting work, one has to know how many seedlings of the different species will be available. This information should be collected every month where nursery production time is under five months, otherwise every two months.

The form opposite should be used to collect and present this information. Where applicable, categories like stumps, cuttings, oversize pots, grafted plants, should be added.

NURSERY INVENTORY FORM

Inventory of ______________________ nursery
(name)

Location ______________________

Person in charge ______________________ Supervisor ____________

Total area ______________________ ha Bed area ____________ ha

Number of workers ______________________

Species	Date: No.	Height	Date: No.	Height	Date: No.	Height	Date: No.	Height
1.								
Sown								
Germinated								
Pricked out								
In pots								
Bare-rooted								
2.								
Sown								
Germinated								
Pricked out								
In pots								
Bare-rooted								
3.								
Sown								
Germinated								
etc.								

APPENDICES

APPENDIX 1: COMPOST MAKING

APPENDIX 2: ADDING USEFUL BACTERIA AND FUNGI TO THE NURSERY SOIL

APPENDIX 3: SOME REMARKS ABOUT CONTAINERS

APPENDIX 4: NOTES ON ORGANISING AND CARRYING OUT TRAINING ON TREE NURSERIES

APPENDIX 1: COMPOST MAKING

Purpose

To produce material rich in nutrients for potting soil mixture, at least partly in the nursery instead of buying manure or artificial fertilisers.

Materials to use

All kinds of organic matter (leaves, grasses, farm manure, chopped left-over seedlings, etc.):

- weeds and grasses should not contain seeds;
- farm manure is very valuable but never to be used fresh (several months of decomposition);
- woody materials (small twigs, left-over seedlings) make good compost, but decompose slowly.

Preparation

Step 1 Place layers of organic waste 20-25 cm thick alternating with layers of topsoil (5 cm thick).

Step 2 Keep moist but not wet throughout its maturation.

Step 3 Turn over when the compost shrinks after heating and arrange a new heap.

Step 4 Repeat turning over several times until compost has matured.

Ready for use

The compost is ready when:

- all organic particles have decomposed;
- the colour has turned to grey-black, the material smells of good soil, not of rot.

1-1.5 m^3 of compost is sufficient for about 10,000 pots.

COMPOST MAKING

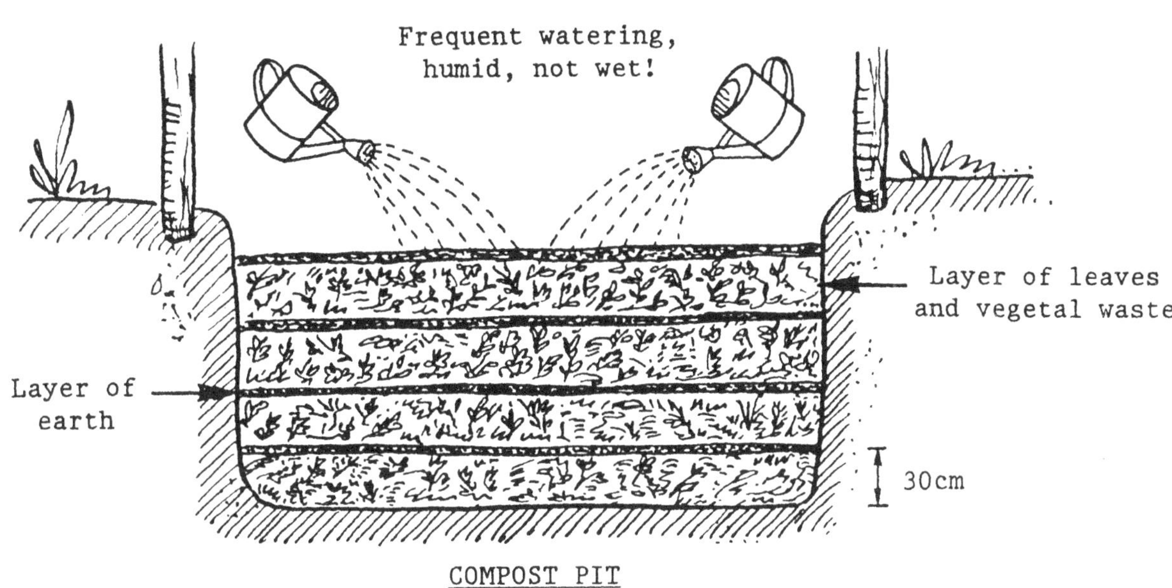

COMPOST PIT

COMPOST MOUND

When the compost shrinks after heating up, turn it over and arrange a new heap

APPENDIX 2: ADDING USEFUL BACTERIA AND FUNGI TO THE NURSERY SOIL

While some bacteria and fungi can cause diseases in seedlings, many others are harmless or even beneficial. Some tree species do not grow well if bacteria or fungi normally associated with their roots are missing. This is particularly true for the bacteria and fungi of leguminous species which form nodules on their roots because the nodules enable the plants to obtain nitrogen from the air. Another example is pines where the fungi called mycorrhiza help to absorb nutrients from the soil.

For leguminous trees, a treatment is only needed if the seedlings grow poorly and do not form nodules or if a new species is introduced. To make sure bacteria or fungi are present, soil from under well-growing trees of the species is mixed with the potting soil or crushed nodules are mixed with the irrigation water.

For pines, soil from under well-established trees can be mixed with the potting soil or some trees with the fungi are left along the beds. Where no plantations of the pine species in question are found close to the nursery, a plot with mycorrhiza mother trees should be planted.

ADDING USEFUL BACTERIA AND FUNGI TO THE NURSERY SOIL

Nodules from leguminous and some other species

Soil from leguminous tree is mixed with potting soil

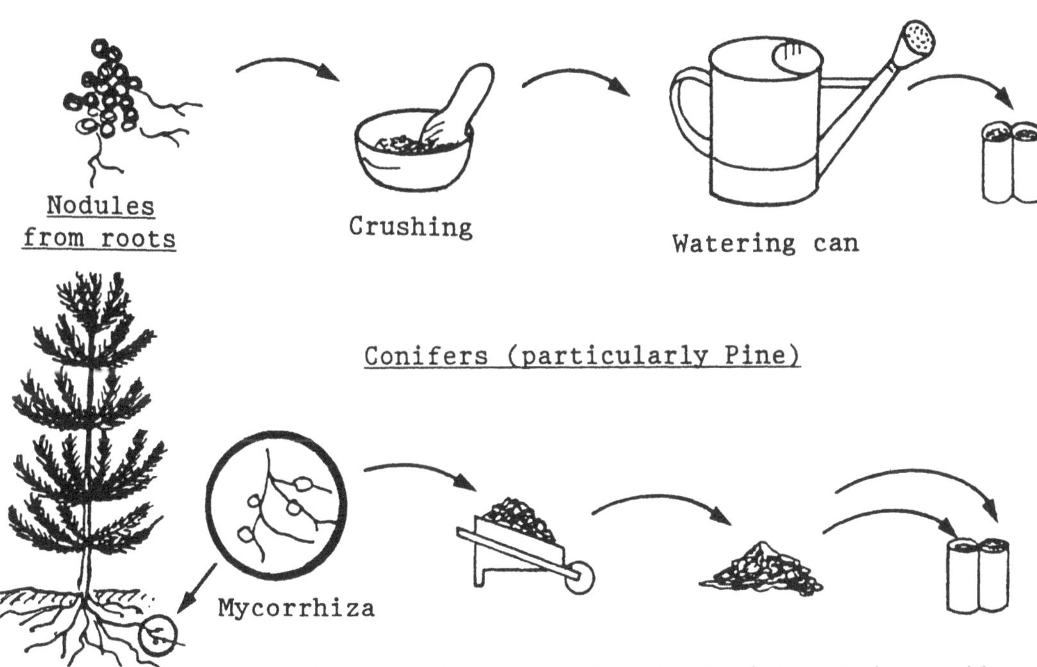

Soil from pine trees is mixed with potting soil

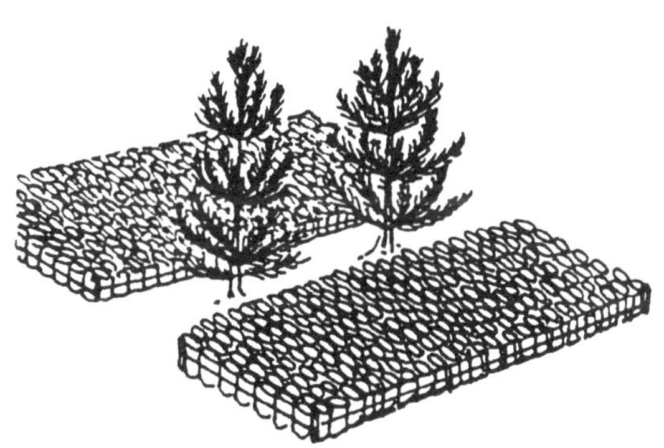

Mycorrhiza mother pines between beds

Production of mycorrhiza soil under pine tree

APPENDIX 3: SOME REMARKS ABOUT CONTAINERS

Plastic pots

Polyethylene tubes are very convenient as potting material and are widely used.

The minimum thickness of the plastic should be 0.02 mm. More commonly, 0.04-0.06 mm material is used.

The pots are cheaper if supplied in the form of rolls of endless tube which has to be cut to the desired length. These pots are open at both ends. Other pots are closed at the bottom like sacks. This has the advantage that potting soil does not fall out during handling and transport. There are, however, two disadvantages of closed pots: they need enough drainage holes otherwise water will stagnate and there is a big risk of roots growing in a spiral inside the bag.

Plastic for pots is either transparent or black. Both are useful. Transparent material can lead to algae growth inside the pot, which is bad for root development. Black pots may heat up too much when exposed to sunlight. Both problems can be reduced through appropriate bed construction as explained in Chapter 1.

The desirable length and width of the pots depends on climate (the drier, the bigger) and transport and planting conditions (bigger containers for difficult sites, non-reliable transport and possible delays in planting).

Some common pot sizes, the space required in beds and the amount of soil for filling are given in the table opposite. When quoting pot diameter, one has to specify whether one refers to flat (empty) ones or round (filled) ones. For orders, flat width and weight are used. The table gives some of the relations.

SOME COMMON POT SIZES, SOIL AND BED AREA REQUIREMENTS

Pot dimensions (cm)			Bed area requirements		Potting requirements		No. of pots per kg (0.04 mm gauge)
Length	Width flat	Diameter filled	Pots per m^2 bed	m^2 per 10,000 pots	Vol/pot (cm^3)	m^3/10,000 pots (rounded)	
18	5.5	3.5	772	13	173	1.8	1 365
18	6.5	4.1	625	16	238	2.4	1 160
18	10.0	6.5	240	42	597	6.0	750
15	10	6.5	240	42	497	5.0	900
18	12.5	8.0	156	64	904	9.0	600
20	12.5	8.0	156	64	1 005	10.0	540
30	12.5	8.0	156	64	1 507	15.0	360
25	25	16.0	39	256	5 024	50.0	216

APPENDIX 3: SOME REMARKS ABOUT CONTAINERS

Other types of containers

Polyethylene tubes may be difficult and costly to obtain. In this case, the possibility of using other potting material should be considered.

Bamboo:

Bamboos with an inner diameter of at least 4 cm can be transformed into containers by cutting bamboo poles by the internodes. The latter becomes the bottom of the container into which a drainage hole is punched with a machete.

Where bamboo with thick walls is used, a cut should be made on one side before filling to make the removal of the container at planting time easier.

Veneer tubes:

Can be made where waste veneer is available. The veneer is cut, wrapped around a wooden roll with the veneer grain running lengthwise and then held in position by stapling the top and bottom end.

Beer and other metal or paper cans:

These are sometimes available in substantial quantities. Metal cans fitted with slots for drainage and easy removal at planting make suitable containers. They have the disadvantage of heating up too much if exposed to direct sunlight. Waxed paper cans used for milk or fruit drinks are not so solid and require more care in handling.

Banana stems

Fibre from banana stems can be used for making transplanting containers of various sizes.

Banana leaves:

The use of banana leaves and similar fibre containers is rather time-consuming and durability of the containers is limited. They may be suitable for fast-growing seedlings or for wrapping up seedlings grown without containers but lifted with the earth ball around the roots.

OTHER TYPES OF CONTAINERS

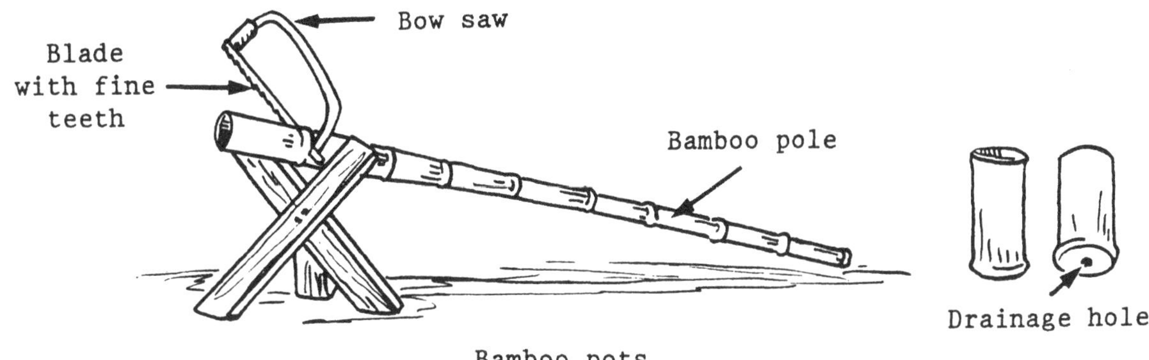

Bamboo pots

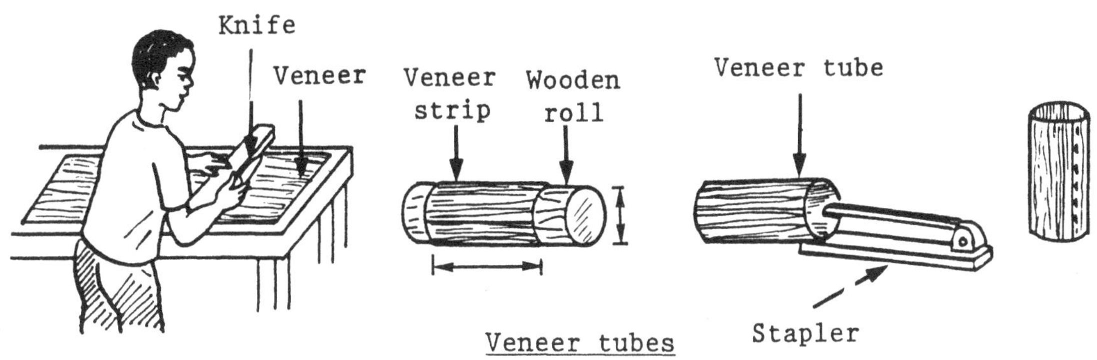

Veneer tubes

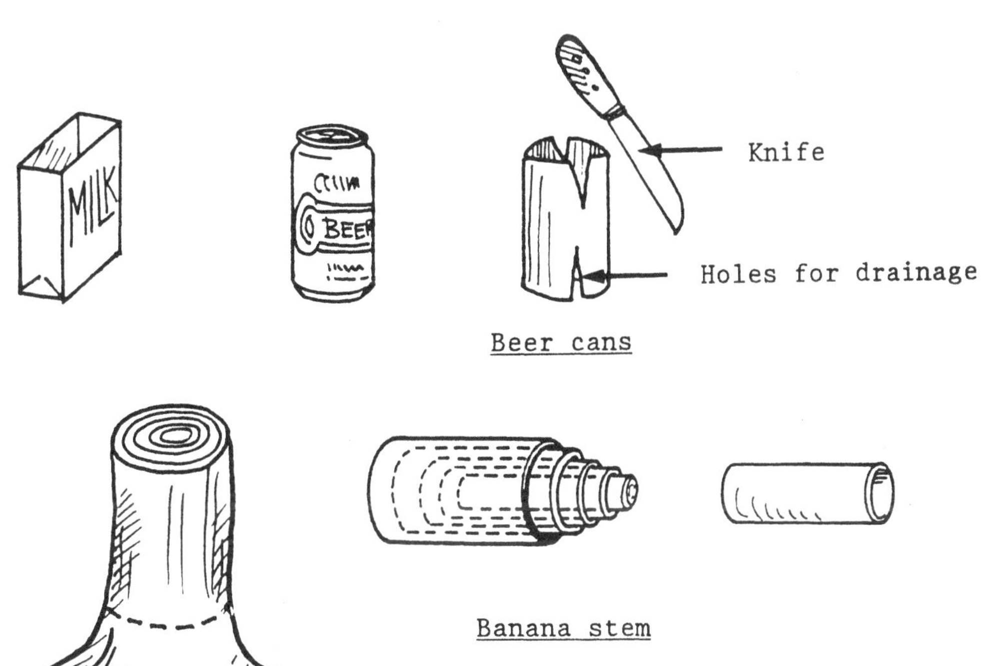

Beer cans

Banana stem

APPENDIX 4: NOTES ON ORGANISING AND CARRYING OUT TRAINING ON TREE NURSERIES

---

1. Introduction

This booklet is concerned with the establishment, operation and administration of small and medium size tree nurseries. The booklet can be used in many ways: as a simple reference manual or field guide for technical personnel - or as a training booklet to be used in the context of a training course. This section is intended to advise a trainer on how to use the booklet during a training course.

The tasks involved in establishment, operation and administration are quite different and they will normally be carried out by different categories of workers. Training sessions should be tailor-made for the particular training needs. Before beginning any training programme, you should ask yourself three simple questions:

1. Why do I need to carry out training?
2. Who needs training?
3. What do they need to be trained to do?

Why do I need to carry out training?

Before undertaking training, you should first make sure that it is necessary. You have to identify tasks which are not performed to the project's satisfaction because those involved lack certain know-how and skills which can be imparted by training.

This booklet therefore needs to be adapted to fit those training needs you have found yourself. Training may need to be more comprehensive if the trainees have never done any nursery work before or more selective if certain activities are new to them or not done in the required way. Leave out those sections which do not apply to your particular situation, and also feel free to include new material which is not contained in this booklet that may be required for your particular training needs.

Who in the project need training?

This booklet can be adapted to fit different categories of trainees, each one concerned with particular aspects of forest nursery work. Before embarking on training, you should have a clear view of who you are going to train. This means knowing not only what particular jobs and tasks your personnel will be required to carry out, but also discovering what your personnel already know and also what their current level of knowledge, basic skills and work experience is.

Generally speaking, the booklet can be adapted for use with either site foresters or nursery foremen, as well as with various categories of intermediate technical personnel.

For project foresters, emphasis will usually be placed on location, and design of nurseries as well as production planning. However, they should also have a thorough understanding of the various tasks of nursery establishment and operation since they will be required to train and supervise their site personnel accordingly.

Nursery foremen will be primarily concerned with nursery operation and to a varying extent with planning, work organisation and recording. Although it may be helpful for foremen to understand on what basis a nursery site was chosen and how a nursery was designed, clearly, these tasks should be left to personnel with appropriate prior qualifications. Your training of foremen therefore should either adapt or omit altogether those sections of the booklet not appropriate.

All the above categories of personnel are likely to have prior theoretical knowledge and/or practical experience.

Finally, while the booklet cannot normally be used for the direct training of workers, foremen should be instructed on how to pass the acquired skill on to them.

What do they need to be trained to do?

Once you have clearly identified your programme's training needs, and the category and level of personnel to be trained, you should determine the specific tasks which will have to be carried out by the trainees in your particular project.

For refresher training, carefully analyse nursery records, observe the execution of nursery work at various stages and ask personnel about difficulties they encounter to identify common errors and gaps in know-how.

2. How will the training be carried out?

Where to train

Training may be carried out in a classroom, in a nearby community centre, in a calm, shaded spot outdoors close to the project worksite or on the job.

For how long?

The duration of the training can vary considerably from a week or more to only a few hours. Courses lasting several days may be appropriate in early phases of the project when a general introduction is needed for new personnel.

Below a plan for a five-day training course is given, covering all tasks presented in this booklet. You should modify this schedule to fit your training needs and the category of personnel you are training. In all likelihood, your training needs will require that more time be given to each training subject - and that the contents of this booklet be divided into several training sessions interspersed with periods of practical work.

APPENDIX 4: NOTES ON ORGANISING AND CARRYING OUT TRAINING ON TREE NURSERIES

Training should include field visits and practical exercises. The trainee should visit at least two nursery sites presenting different characteristics.

Structuring a lesson

First of all spell out precisely what the trainees are supposed to know and to be able to do after the lesson has been given. In other words, what is the objective of the lesson? The course programme proposed below provides some examples of objectives. For each major task there should be a sequence of explanation, demonstration, practical exercises by each trainee, correction of execution through instructors and continuation of exercise to perfection skills.

To facilitate understanding and learning of more complicated operations they are best broken down into a sequence of individual steps. An example is given in Chapter 3 for the pricking-out operation.

Avoid overloading the programme, which would force you to rush sessions. Never skip practical exercises.

Preparing a session

When planning courses and sessions take into account that nursery work is seasonal. Many activities can only be shown and practised meaningfully at certain times (e.g. root pruning, grading of seedlings, grafting). For others, some preparatory work is needed if they are to be taught during the same course (e.g. sowing and pricking-out requires presowing a batch to have reached right size during course).

Trainees are much more likely to remember how a job has to be done if they actually carried it out during the session rather than listening to explanations only. For this reason, include carefully planned practical exercises for all important tasks.

Training materials

For the explanation part, a blackboard, board dusters and coloured chalk should be used. Where available, flipcharts and supports are very helpful because more elaborate drawings can be prepared in advance and used repeatedly.

Tools, instruments, recording forms and supplies have to be available in sufficient quantity to permit individual practice for each participant or work in small groups.

Collect specimens of good and bad tools, seed, seedlings, potting soil, shading mats, etc. to illustrate your explanations.

Provide all trainees with a notebook as well as pen and pencil.

SUGGESTED INTRODUCTORY COURSE (ONE WEEK)

1st day

Subject: Description of a nursery (Chapter 1)

Objective(s): At the end of this session, the trainee should be able to:

- identify and describe the basic facilities of a nursery;
- demarcate the location of seed- and pot-beds and structures on the site with measuring tape, tracing line and pegs as indicated on a sketch;
- to choose the appropriate type of beds, fence, shade and water supply for the size and locality of the nursery.

Schedule Morning: training session presentation and discussion of existing nursery site.
Afternoon: visit to new nursery site, demarcation and selection exercises.

2nd day

Subject: Seeds and potting soil (Chapters 2 and 3.1-3.3).

Objective(s) At the end of this lesson, the trainee should be able to:

- explain the importance of seed provenance and name mother tree selection criteria for different planting purposes;
- carry out a calculation of seed requirements;
- explain the principles of seed extraction and storage;
- describe the characteristics of a suitable soil mixture and ways to produce it;
- demonstrate correct filling of pots with the help of scoop and/or funnel.

Schedule: Morning: presentation, exercises calculation, excursion in vicinity, appraisal of seed and potting soil sources.
Afternoon: at nursery demonstration seed extraction and pot filling, exercises.

APPENDIX 4: NOTES ON ORGANISING AND CARRYING OUT TRAINING ON TREE NURSERIES

3rd day

Subject: Sowing, care, tending and preparations for planting out (Chapter 3.4-3.7).

Objective(s): At the end of this lesson, the trainee should be able to:

- place pots correctly in beds;
- pretreat seed, take a decision on direct or in-pot sowing;
- prepare a seedbed and sow seed of different size;
- determine the right moment for pricking-out and demonstrate this operation;
- water, tend and shade seed- and pot-beds correctly and explain principles of pest and disease prevention and control;
- explain grading of seedlings of different quality, the concept of hardening-off and preparations for transport;
- identify, describe and use suitable tools for each operation.

Schedule: Morning: training session and discussion.
Afternoon: demonstrations and practical exercises in existing nurseries.

4th day

Subject: Bare-rooted seedlings and stumps (Chapters 4 and 5)
(In projects where the production of bare-rooted seedlings or stumps is irrelevant, the content of day 3 should be spread over 2 days).

Objective(s): At the end of this lesson, the trainee should be able to:

- explain the advantages and disadvantages of bare-rooted seedlings and stumps compared to potted stock;
- to cultivate beds for bare-rooted plants and to transplant;
- to root-prune, lift and pack bare-rooted stock;
- to prepare cuttings.

Schedule: Morning: Presentation, demonstration and discussion.
Afternoon: practical exercises.

5th day

Subject: Nursery planning, work organisation and recording (Chapter 7).

Objective(s): At the end of this lesson, the trainee should be able to:
- explain the use of a production calendar and calculate sowing dates;
- describe and fill out the nursery recording forms;
- explain principles of work organisation, choice of workers, remuneration and working conditions.

Schedule Morning: Presentation, demonstration and discussion, exercises production calendar.
Afternoon: Exercises nursery inventory, form filling and working conditions (discussion of possible improvements)
Winding-up of course, evaluation discussion.

Note: Where grafting of fruit trees is a relevant operation, try to obtain the collaboration of an experienced horticulturalist at least for the demonstrations and practical exercises.

GENERAL ADVICE TO TRAINERS IN CARRYING OUT TRAINING

ENCOURAGE ACTIVE PARTICIPATION BY TRAINEES:
DON'T LECTURE !

There is sometimes a tendency in training courses for the trainer to act like a teacher in school, and to read or lecture from training booklets. This tendency should be avoided. Trainees will remember information better if they participate actively in discussion and if there is a free exchange of views and of questions between everyone involved in the training course. Ensure that the atmosphere is sufficiently relaxed. Remember that often there will be no strict "right or wrong" answer to a question.

Equal attention should be paid to each speaker. You should listen attentively and let him understand that his ideas and opinions are both interesting and important. It is sometimes useful to take note of participants' suggestions while they are speaking, jotting them down on a flipchart or blackboard. A summary of these notes may prove useful for later discussions or serve as a starting point for the documentation of local nursery experience.

APPENDIX 4: NOTES ON ORGANISING AND CARRYING OUT TRAINING ON TREE NURSERIES

GUIDING THE DISCUSSION

There are times during a discussion when everyone wants to speak at the same time. Should this situation arise, the trainer must insist that the group listento one person at a time. If one speaker holds the floor for too long, the trainer should interrupt him, reminding him that his colleagues would also like to speak. If the discussion drifts away from the aim of the discussion, briefly recall the main points. Emphasise important points raised by repeating them.

EVALUATING YOUR TRAINING COURSE

Make sure that trainees have really learned how to carry out the jobs you are training them for. For this reason, you should frequently evaluate how effective training has been. This evaluation can be done in different ways.

- Sometimes, you may simply ask questions in order to be sure the presentation has been understood.

- The best way to evaluate training is by asking trainees to carry out a particular task required.

- Remember, the real test of whether the training course has been worthwhile or not is the trainees' ability to do their jobs at the end of the course - and not their ability to recite an answer to any particular question.

- Evaluation should be carried out frequently, in the middle as well as at the end of each training session. In this way, you will avoid beginning a new subject before trainees have learned the material already presented. Be sure all of your trainees are "with you" before beginning a new subject.

- In evaluating your training, you should refer to the objectives of each session. You may use the objectives outlined in the suggested training programme above, or you may draw up your own training objectives.

SOME USEFUL GUIDES/HANDBOOKS

1. Weber, F.R.: Reforestation in arid lands, 2nd ed., 1986 (includes descriptions of some useful species in Sahelian Africa), 335 pp., ISBN 0-86619-264-6. Available from: VITA, 1815 North Lynn St., Suite 200, Arlington, VA 222 09, USA..

2. GTZ: Manual of reforestation and erosion control for the Philippines, 1976, 569 p. Available from: German Agency for Technical Co-operation (GTZ), P.O. Box 5180, D-6236 ESCHBORN, Fed. Rep. of Germany, ISBN 3-88085-020-8

3. Manual de viveros forestales en la Sierra Peruana, Proyecto FAO/Holanda/INFOR, 2nd edition, 1985, 123 pp. Available from: Cahbide 805-Piso 7°, Jesús María, Apartado Postal 11016, LIMA 14, Peru.

4. Manual de viveros y plantaciones escolares, Proyecto FAO/Holanda/INFOR, 1985, 51 pp. Available from above address.

5. SIDA/Swedforest: Village nurseries, 1984, 37 pp. Available from: Swedish International Development Agency (SIDA), S-105 25 STOCKHOLM, Sweden.

6. Nepal-Australia Forestry Project, Kathmandu, Nepal: Raising eucalyptus seedlings and establishing eucalyptus plantations in the hills of Nepal, Technical note 1/74, 2nd edition, 1983, 18 pp.;

7. Nepal-Australia Forestry Project, Kathmandu, Nepal: Establishing and running forest nurseries in Nepal, 2nd edition, 1983, 47 pp.;

8. Nepal-Australia Forestry Project, Kathmandu, Nepal: Plant propagation for reforestation in Nepal, 1983, 104 pp. 6, 7 and 8 available from Nepal-Australia Forestry Project, GPO Box, KATHMANDU, Nepal.

9. Projet pilote forestier (PPF), Kibuye, Rwanda: Création et conduite d'une pépinière forestière, Fiche technique no. 1, 2nd edition, 1976, mimeographed series.

10. Ethio-German Afforestation Project, Addis Ababa, Ethiopia: Forest nurseries, Project Working Paper No. 5, undated, mimeographed leaflet series.

11. Zimbabwe Forestry Commission: Planting and managing woodlots, Booklet No. 2, 23 pp., undated. Available from: Publicity and Extension Officer, Rural Afforestation Division, P.O. Box HG 139 HIGLANDS, HARARE, Zimbabwe.

REFERENCES

---

12. Evans, J.: Plantation forestry in the tropics, Oxford University Press, Oxford, United Kingdom, paperback edition, 1984, 471 pp.. ISBN 0-19-85948953.

13. Turnbull, John (ed.): Multipurpose Australian trees and shrubs, 1986. Available from: Australian Centre for International Research, GPO Box 1571, CANBERRA, ACT 2601, Australia.

www.ingramcontent.com/pod-product-compliance
Ingram Content Group UK Ltd.
Pitfield, Milton Keynes, MK11 3LW, UK
UKHW061817190726
13853UKWH00006B/2198